The use of coatings in industry is growing and will continue to grow
because of the economic and technical advantages they offer over
uncoated materials. Although a wide variety of materials and application
techniques are available, much less is known about the properties of
specific coatings and their measurement. This volume contains some
twenty-six papers that were presented at a symposium organised to
explore these questions and they represent state-of-the-art technology.
The symposium was divided into five sessions dealing with new coating
technologies, measurement of coating properties, marine coatings, field
applied coatings for corrosion control, and tribological coatings.

Mechanical Properties, Performance, and Failure Modes of Coatings

Mechanical Properties, Performance, and Failure Modes of Coatings

Proceedings of the 37th Meeting of the
Mechanical Failures Prevention Group,
National Bureau of Standards,
Gaithersburg, Maryland,
May 10-12, 1983

Edited by

T. Robert Shives
National Bureau of Standards
Gaithersburg, Maryland

Marshall B. Peterson
Wear Sciences, Incorporated
Arnold, Maryland

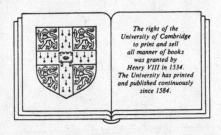
CAMBRIDGE UNIVERSITY PRESS

Cambridge

London ▪ New York ▪ New Rochelle ▪ Melbourne ▪ Sydney

CAMBRIDGE UNIVERSITY PRESS
Cambridge, New York, Melbourne, Madrid, Cape Town, Singapore, São Paulo, Delhi

Cambridge University Press
The Edinburgh Building, Cambridge CB2 8RU, UK

Published in the United States of America by Cambridge University Press, New York

www.cambridge.org
Information on this title: www.cambridge.org/9780521103558

First published 1984
This digitally printed version 2009

A catalogue record for this publication is available from the British Library

Library of Congress Cataloguing in Publication data

Mechanical Failures Prevention Group.   Meeting (37th :
    1983 : National Bureau of Standards)
    Mechanical properties, performance, and failure modes
of coatings.

    1. Coatings--Congresses.   I. Shives, T. R.
II. Peterson, M. B. (Marshall B.)   III. Title.
TA418.9.C57M43   1983   667'.9   83-26136

ISBN 978-0-521-26420-4 hardback
ISBN 978-0-521-10355-8 paperback

Table of Contents

PREFACE

The 35th Meeting of the Mechanical Failures Prevention Group (MFPG) was held May 10-12, 1983, at the National Bureau of Standards in Gaithersburg, Maryland. The program, which focused on the mechanical properties of coatings and their measurement, was organized by the MFPG Materials Durability Evaluation Committee under the chairmanship of Marshall Peterson of Wear Sciences, Inc. Appreciation is expressed to the committee, the session chairman, and the contributors for an excellent program. Appreciation is also expressed to the National Bureau of Standards, the Office of Naval Research, and the Naval Air Systems Command for financial support.

Special thanks are due to Sara R. Torrence of the NBS Public Information Division for the meeting, hotel, and social function arrangements, and to Kathy C. Stang for handling financial matters.

The Mechanical Failures Prevention Group (MFPG) was organized in 1967 to stimulate cooperation among various segments of the engineering and scientific communities in an effort to reduce the incidence of mechanical failures and to develop methods to predict mechanical failures. The MFPG is an interdisciplinary group with a strong application orientation. The membership includes professional personnel representing a wide variety of scientific and engineering disciplines. Individual members are associated with government agencies, industry, universities and research institutes. The MFPG also cooperates with appropriate committees or units of professional societies.

Typically, there are two MFPG symposia each year. Responsibility for the technical program for these symposia rotates among the four MFPG technical committees: 1) Mechanisms of Failure, 2) Machinery Durability, 3) Detection, Diagnosis and Prognosis, and 4) Materials Durability Evaluation.

T. Robert Shives
Executive Secretary, MFPG

SESSION I

NEW COATING TECHNOLOGIES

CHAIRMAN: W. WINER
           GEORGIA INSTITUTE OF TECHNOLOGY

A REVIEW OF COATING TECHNOLOGY

S. Ramalingam
University of Minnesota
Minneapolis, Minnesota 55455

Abstract: Friction- wear-induced mechanical failures may be controlled to extend the life of tribological components through the interposition of selected solid materials between contacting surfaces. Thin solid films of soft and hard materials are appropriate to lower friction and enhance the wear resistance of precision tribo-elements. A variety of thin film coating technologies have been developed to deposit thin solid layers to reduce friction or to extend wear life. They include chemical vapor deposition (CVD) and physical vapor deposition (PVD) processes.

A particular advantage offered by the newer coating processes is that they allow close control of coating thickness and composition not possible with many of the older and traditional processes. Moreover, they allow well-bonded thin films to be deposited at low temperatures. Hence, specific modification of friction and wear characteristics is possible as a finishing treatment requiring no further processing.

In this review, the coating technologies that have been developed and brought into commercial use in recent years are discussed. The specific advantages offered and the benefits realized are identified. Particular attention is directed at coating technologies which qualify as finishing treatments for the modification of the tribological characteristics.

Introduction: A number of new technologies have been developed in recent years for the modification of surface properties that offer considerable flexibility and process economies in surface property modification for tribological control. They rely on the deposition of a range of soft and hard compounds as well as metals. The real advantage offered by the new technologies is that they allow close control of coating thickness and composition not available in most standard treatments. They also permit combinations of coatings - for example, a soft overlay on a hard coat - in such a manner that no further finishing of the tribo-element is necessary. These processes include pyrolytic decomposition, chemical vapor deposition, physical vapor deposition, reactive evaporation, sputtering, arc coating, etc.

In this review, surface coating techniques particularly suited for the treatment of precision tribological elements are discussed. This review is not intended to be comprehensive. Key concepts are referred to and technical details are left out. The reader should consult the literature cited for a more complete understanding of the specific processes discussed.

Relevant Coating Techniques: Pyrolytic decomposition of organo-metallics, halide metallurgy, electrosynthesis in molten salt bath and a number of electrically-assisted vacuum coatings are among the principal surface coating techniques now available to improve tribological characteristics of mechanical components. Each of these possesses one or more special characteristics not readily available in another. Hence, all of the techniques will have to be considered before selecting a particulr technology for a specific tribological application. CVD (halide metallurgy) and PVD (vacuum coating) techniques are presently the two widely used

3

processes for surface coating to protect tribological components.

Pyrolytic Decomposition: In some organo-metallic compounds, the metal atom can be in a valence state which permits easy decomposition of the compound by thermal means. By heating the substrate to be coated in a flowing mixture of inert gases and appropriate organo-metallics, thin coats of chromium (1), aluminum (2), gold (3), silver (4), nickel (5), etc., are easily produced. Gold and silver coatings so produced can be useful for soft coating of tribo-elements.

Temperatures involved in metallic coating are usually low. The coating rates are also low. Thin noble metal soft coats can therefore be produced without serious thermal problems to afford tribological protection to precision engineering components.

Hard coating, especially with oxides, is also possible by the use of pyrolytic decomposition. Alumina, silica, zirconia, and thoria coatings may be deposited by the thermal decomposition of metal alkoxides (6). The temperatures involved are, however, higher than those usually encountered in the deposition of metal layers. Production of alumina coating by the decomposition of aluminum isopropoxide, for example, requires (6) temperatures of 700-800° C. Hard coating by thermal decomposition may, hence, require a thermal treatment following thin film deposition.

Decomposition of nickel carbonyls (5) is now in use to construct foundry patterns to withstand abrasive wear. Soft, noble metal coats produced by thermal decomposition are also in use. Apart from this, pyrolytic decomposition has not found wider use in tribological applications.

Chemical Vapor Deposition: In CVD processes, halide compounds are reacted with hydrogen in an appropriately designed chemical reactor to deposit thin films of needed metals or compunds on hot substrates. The halide compounds needed may also be produced by reacting hot metal chips with selected halogens within suitable containers. To deposit carbides, nitrides, oxides, borides, etc., the wear-resistant coatings of interest to tribologists, the chemical potential of the needed atomic species is enriched by entraining suitable source gases in the process streams. Soft coat deposition by CVD is not common.

A schematic representation of CVD facilities presently in use is shown in Figure 1. Some of the characteristic reactions exploited for CVD coating of different materials are shown in Table 1.

The coating temperature in CVD is determined by the energetics of the reaction chosen. Free energy change accompanying the selected reaction must be negative at the desired coating temperature. To deposit TiN, for example, the following reactions may be used:

$$TiCl_4 + 1/2 \ N_2 + 2H_2 \rightleftharpoons TiN + 4 \ HCl$$

$$TiCl_4 + NH_3 \rightleftharpoons TiN + 3 \ HCl + 1/2 \ Cl_2$$

$$TiCl_4 + NH_3 + 1/2 \ H_2 \rightleftharpoons TiN + 4 \ HCl$$

4

TABLE 1: CHEMICAL VAPOR DEPOSITION

REACTION TYPE	EXAMPLES
Pyrolysis	$Ni(CO)_4(g) \rightleftharpoons Ni(s)^* + 4CO(g)^*$
Reduction	$SiCl_4(g) + 2H_2(g) \rightleftharpoons Si(s) + 4HCl(g)$
Oxidation	$SiH_4(g) + O_2(g) \rightleftharpoons Si(s) + 2H_2O(g)$
Hydrolysis	$Al_2Dl_6(g) + 3CO_2(g) + 3H_2(g) \rightleftharpoons$
	$Al_2O_3(s) + 6HCl(g) + 3CO(g)$
Nitride formation	$3SiH_4(g) + 4NH_3(g) \rightleftharpoons Si_3N_4(s) + 12H_2$
Carbide formation	$TiCl_4(g) + CH_4(g) \rightleftharpoons TiC(s) + 4HCl(g)$
Disproportination	$2GeI_2(g) \rightleftharpoons Ge(s) + GeI_4(g)$
Synthesis	$(Ch_3)_2Cd(g) + H_2Se(g) \rightleftharpoons CdSe(s) +$
	$2CH_4(g)$

* (g) = gas; (s) = solid

Equilibrium calculations (7) show that the first two reactions require temperatures in excess of 1000 K for TiN deposition, while the third reaction is possible at temperatures of the order of 600 K.

Since well-bonded coatings are essential for tribological applications, reaction temperature alone is not the sole criterion in the choice of preferred deposition reactions. It may be necessary to deposit an interlayer to enhance coating adhesion, to serve as a diffusion barrier, etc. Over such a layer, low temperature depositions can be made to produce dense, fine grained coats with excellent wear properties (7).

In CVD, the deposition of a specific solid phase is usually desired. Process streams and operating conditions will have to be chosen to ensure process repeatability, to obtaindesired coating rates, to maximize deposition efficiency, etc. To attain these goals, CVD phase diagrams (8) and computation packages have been developed (9) in order to identify the key process control variables. A CVD phase diagram for TiB_2 deposition taken from Reference 8 is shown in Figure 2. A similar diagram for B_4C deposition, also from Reference 8, is shown in Figure 3. The CVD phase diagram for tungsten carbide deposition may be found in Reference 10. From Figures 2 and 3, it is clear that while titanium diboride deposition is possible over a range of conditions, deposition of boron carbide requires close process control. A similar situation is encountered in the deposition of tungsten monocarbide (10).

Equilibrium computations which allow these diagrams to be calculated enable the determination of specific process variables such as temperature, total pressure, ratios of chemical elements in the input gas, etc., in order to identify the condensed phase which will deposit, the equilibrium deposition rate, and the deposition efficiency of reactions. Gas phase reactions and resulting porous coatings can then be avoided to produce dense coatings with excellent tribological characteristics. It should be recognized that kinetic variables such as the total flow rate, coating system geometry, substrate chemistry, and surface finish also play a role in determining the quality of thin coats deposited. Uniform coating coverage of complex part geometry without the use of specialized coating fixtures is possible with CVD. This is not readily accomplished with most PVD processes.

Taking the cited factors into consideration and recognizing that the deposited films may interact diffusively with the substrate, optimal conditions for the deposition of tribological coatings can be identified. When adverse diffusive reactions are unavoidable, a barrier coating, also deposited with CVD, may be used. Sequential or multi-layer coatings are clearly possible through the change of process streams and deposition temperatures.

Chemical vapor deposition is presently an established technology with a sound scientific base. Cobalt bonded carbide tool inserts are now routinely CVD-coated with TiC, TiN, HfN, Al_2O_3, etc., singly or in combination. Major benefits in wear resistance have been demonstrated in commercial use.

Despite wide spread use in tooling applications, CVD has had a far smaller impact in general tribological applications. This is in part due to the higher temperatures encountered in current CVD practice, 800° to 1000° C, for the deposition of hard compound films. In the case of tribo-elements made of steel, the high CVD temperatures used necessarily require post processing following CVD coating. Coating internal stresses and thermal stress induced cracking of the deposited films is then common. A refinishing of the CVD-coated precision components is further necessary, as the coated parts usually do not possess satisfactory, as-coated, surface finish. Distortions due to thermal treatment following CVD will also have to be removed.

In spite of these limitations, well-bonded CVD coatings, especially the TiN coatings on steel, hold great promise in severe tribological applications. Significant scuff resistance exhibited by CVD-coated steel components (11) suggests, that in hostile environments, i.e., higher temperature applications where effective liquid lubrication is not possible, in sealed transmission systems, in solid lubricated rolling element bearings, and in high vacuum bearing applications (12), CVD-coated tribo-elements can provide superior tribological life. Systematic evaluation of coated elements including gear contacts are essential to validate the expected superior tribological characteristics of CVD coated components.

Electrosynthesis in Salt Baths: High temperature salt bath surface treatments driven solely by differences in chemical potential are well known and form the basis of cyaniding, carbo-nitriding and other similar processes. Attainable surface hardnesses with these processes are low compared with those possible by the deposition of refractory metal compounds. If well bonded, continuous, single phase compounds can be deposited at low temperatures, they can be of considerable utility for friction and wear control. Deposition of such compounds at low temperatures is particularly desirable to avoid post deposition thermal treatments. Electrosynthesis offers a means of depositing the needed hard compound layers.

Early electrosynthesis work (13) demonstrated the feasibility of producing iron, tungsten, and molybdenum carbides in alkali carbonate/borate baths at temperatures which compare favorably with those used in CVD. Current interest in hard layer deposition has led to the revival of salt bath electrosynthesis and deposition research. Within the last year, investigators have demonstrated low temperature deposition of nickel boride (14), synthesis of tantalum and vanadium monocarbides (15), and the electrolytic deposition of tantalum carbide (16).

NRL investigators (16) have identified the deposition conditions for the production of well-bonded layers of Ta_2C and TaC on nickel substrates at temperatures between 750 and 800° C. Preliminary friction and wear tests show promise. Other investigators have shown that nickel boride can be electrodeposited at temperatures between 500 and 670° C. Coatings as thick as 46 microns have been produced in approximately 24 hours. High temperature alloys such as Hastelloy B have been deposited with boride films 6 microns thick at low temperatures.

Since it is known empirically that thick wear resistant layers are not necessary to afford significant tribological protection, the demonstrated ability to produce well-bonded thin films at low temperatures suggests that molten salt electrosynthesis is a promising means of hard coat deposition. As in the case of CVD, uniform coating of parts of complex geometry is feasible with salt bath electrosynthesis. Reported coating of substrates with tungsten carbide (15) is further evidence of the potential held by electrosynthesis.

Although the coatings produced apparently possess good film-to-substrate bond strength, the deposited coatings require refinishing before use. Coatings produced are crystalline and are hard. They are likely to possess a columnar structure. Further tribological characterization is essential before coatings produced can be considered to be suitable for engineering applications.

Physical Vapor Deposition Technologies: As noted earlier, the CVD techniques presently available for hard coat deposition are high temperature processes. Electrosynthesis techniques, though promising, are not yet perfected. Since hard and soft coating of ferrous components require low coating temperatures, i.e., temperatures not exceeding those used for tempering following hardening, one will have to resort to one or more of the electrically assisted vacuum coating processes to produce the needed thin films for tribological protection.

A variety of electrically assisted, thin film deposition techniques have been developed to produce the needed coatings. They include ion plating, activated reactive evaporation, DC and RF sputtering, arc coating, coating with plasma discharge, etc. Vapor species produced by thermal evaporation or ion bombardment are used in these processes to obtain thin coats of soft and hard materials. Virtually all of the known hard materials, soft materials, and a number of nonequilibrium structures can be produced with one or another of the electrically assisted coating techniques. When coatings of compounds are needed, the vapor species produced may also be reacted with process gases to concurrently synthesize and deposit needed materials. The principal vacuum coating techniques suitable for tribological coating of precision mechanical components are briefly described and discussed in the following paragraphs.

A schematic of a simple vapor deposition system for thin film deposition is shown in Figure 4. It consists of a vacuum chamber, a melt power supply, a filament, boat or cooled crucible assembly to contain the melt pool, and substrate holders. When substrate heating is permissible or necessary, appropriate means are provided in the substrate holder for sample heating. Simple thermal vapor deposition may be carried out with wire filaments or resistance heated baskets. For granular or powder feed-stocks, resistance heated bare metal boats of W, Mo or Ta, fabricated from sheets are satisfactory. For large scale coating, electron beam melting and induction melting may be used.

In thermal vapor deposition, the mean free path of the coating flux is large and the vapor species arrive at the sample surface without extensive gas collision. Coating is a line-of-sight process. Coating rate depends on the melt power and is inversely proportional to the square of the source-to-sample distance.

The mean energy of the coating flux produced by thermal evaporation is $3/2$ kT^V, where k is the Boltzmann constant and T^V is the average vapor temperature. The mean energy of the vapor flux is insufficient to displace adsorbed species and contaminants on surfaces to be coated. Coatings produced therefore usually exhibit poor film-to-substrate adhesion. This problem can be remedied by injecting electrical charges into the vapor flux with a 'probe electrode' as shown in Figure 4. The substrate can also be biased with a positive potential. Ionized coating flux is then accelerated to impinge on the substrate with greater kinetic energy than in simple thermal vapor deposition to produce well-bonded films (17).

The structure, properties, and adhesion of the films deposited are affected by the substrate temperature permitted during film deposition. Textured films are produced at low substrate temperatures. Columnar structure is obtained at intermediate temperatures. At homologous temperatures above 0.4 to 0.5, equiaxed, but coarse, grains are produced. The original surface finish of the substrate is not duplicated by this coating technique.

By admitting reactive gases during film deposition, coatings of compounds are obtained. Control of compound stoichiometry requires coupling of gas admission rate with the evaporation rate. Dynamic pumping is the common practice. Physical vapor deposition processes with E-beam melting and probe electrodes for injecting charges into the coating flux is reffered to as Activated Reactive Evaporation (17).

Ion Plating, illustrated in Figure 5, is a variant of thermal vapor deposition. Vapor flux is directed at the substrate maintained at a high negative potential. System pressure and deposition conditions are adjusted such that the substrate is enveloped by a glow discharge plasma during thin film deposition. Due to the glow discharge, material is continuously sputtered from the substrate. By continuously cleaning the substrate with sputtering as it is being coated, well-bonded films are laid. Improved film adhesion is obtained.

Use of inductively melted vapor sources allows high coating rates. In ion plating, the use of E-beam melting is not straight-forward, since ion plating pressures are not compatible with good E-beam operation practice. With differential pumping, this difficulty can be overcome. In the normal operating pressures for ion plating, vapor species undergo multiple collisions before they arrive at the substrate. Coating is hence not a line-of-sight process. Fairly uniform coatings and step coverage are therefore possible. Partial replacement of the inert gas used in ion plating with a reactive gas allows reactive ion plating to obtain coating of compounds.

When a high DC voltage is applied to a cathode within a chamber maintained at low pressures, electrons emitted from the cathode ionize the gaseous atoms in the environment. Self sustaining glow discharge can be obtained in such an arrangement by adjusting the chamber pressure and the applied voltage. The positive ions produced in the glow discharge impinge on the cathode and displace the cathode atoms. The sputtered atoms diffuse through the gaseous environment and are deposited on the substrate to be coated. Over a period of time, thin films are produced. When insulating material coatings are to be produced, as the cathode is not an electrical conductor, radio frequency power supplies are essential. Common arrangements for DC and RF Sputtering are shown in Figures 6 and 7.

In sputtering, the coating rates obtained depend on (a) mass and energy of ions impinging on the cathode, (b) target to substrate distance, (c) sputtering pressure, (d) gas temperature, (e) cathode surface condition, (f) cathode geometry and dimensions, (g) shielding, and (h) target materials. Under commonly used coating conditions, the coating rates obtained are a few nanometers per minute.

Sputter coating depends on a momentum transfer process. Sputtered species have velocities nearly an order of magnitude larger than those produced by thermal evaporation. Good film-to-substrate adhesion is hence obtained on well cleaned substrates despite system operation at pressures greater than those in thermal evaporation. At common operating pressures, since multiple collision in the ambient occurs, coating is not a line-of-sight process. Good step coverage and coating uniformity are obtained without much difficulty. Suitable fixturing and drive assemblies permit a wide range of complex component surfaces to be coated with adequate uniformity. Multi-cathode coating systems allow sequential coating of several layers without breaking the vacuum. This is advantageous to obtain well bonded, multi-layer coatings.

To obtain higher coating rates than those possible with conventional DC and RF sputtering, and to coat at lower system pressures in order to obtain well bonded coatings, Magnetron Sputtering has been developed. In magnetron sputtering, crossed electric and magnetic fields are used to greatly increase the ionization efficiency of the sputtering system without raising the system pressure. The magnetic field superimposed on the electric field generates an efficient electron trap which results in the formation of intense glow discharge plasmas confined to the vicinity of the cathode. It allows the plasma to be supported at low voltages and low pressures. "Cold sputtering" at high rates in a good vacuum is facilitated. Apart from the magnetron head, the magnetron sputtering system is identical to the conventional DC sputtering system.

At the operating pressures normally used, magnetron sputtering is a line-of-sight coating process. Coating rates obtained increase linearly with the power dissipation at the cathode. Deposition rates of several hundred nanometers per minute are easily obtained when the cathode power dissipation is several kilowatts.

With the use of appropriately designed anodes and floated substrates, thermal loading on the substrate can be minimized (18) during magnetron sputtering. Experimental studies of temperature rise during high rate sputtering show that nearly half of the thermal loading during coating is due to condensation of the coating flux on the substrate coated. The remainder can only be accounted for by the kinetic energy of ionized species arriving at the substrate. In the case of aluminum sputtering, the mean energy of the coating species has been estimated and found to be several times the binding energy (18). Similar values apparently hold for magnetron sputtering of other materials. The high energy of the coating flux and the use of good vacuum are believed to be responsible for the well bonded films produced with magnetron sputtering.

Partial replacement of the sputtering gas with a reactive gas enables reactive sputtering to produce coatings of compounds. with commercial purity titanium targets, thin TiN films suitable for tribological applications have been successfully deposited (19-21). Coating rates obtained are lower than those for the sputtering of metals. Magnetron heads can be supported by DC as well as RF power supplies. Both DC and RF sputtered TiN coats have been produced for tribological applications.

Arc Coating Technology: Metallic and non-metallic coatings suitable for tribological applications may also be produced with Arc Coating techniques. This technology differs fundamentally from the electrically-assisted coating processes discussed so far. A 'hot', i.e., a thermal plasma, is used with high ion temperatures. Estimates of ion temperatures range up to 10^5 K. In arc coating, a vacuum arc is used for film deposition. A vacuum arc is a low voltage, high current arc sustained with a metal plasma. A feed gas, as in sputtering, is not essential, but may be used to concurrently synthesize and deposit coatings of metal compounds.

An arc coating system consists of a main cathode of target material, an anode and an igniting electrode all maintained in a vacuum of 10^{-6} to 10^{-3} torr. The system is equipped with a low voltage, high current DC power supply both to ignite the arc and to maintain the primary arc between the cathode and the anode (22). A variety of techniques may be used to ignite the arc, and once the arc is ignited, the igniter is deactivated. The primary arc is then sustained between the main cathode and the anode.

In the cited reference (22), an electro-mechanical assembly consisting of a solenoid and an armature, where the armature carries the igniter electrode, is used. Initially the igniter electrode is in contact with the main cathode. With the current passing through the cathode-igniter circuit, the igniter is withdrawn. As the igniter electrode is withdrawn, the arc gap and length increase until arcing is initiated between the main cathode and the anode.

In arcs, the cathode temperature is high since the cathode spot is of a small size and very large current densities, 10^5 to 10^7 amp/cm^2, occur. Severe localized ohmic heating and evaporation ensues resulting in the erosion of cathodic material. Vapor pressure of the cathodic material rises and conduction through the metal plasma results in arc discharge.

Arcs can be sustained over a range of pressures. At the upper end of the pressure range, multiple collisions occur within the coating flux. Coating uniformity and step coverage is hence facilitated. Since coating is not line-of-sight, use of complex fixturing is not necessary.

It is known from arc physics that arcs are of millisecond duration with current levels of several hundred to several thousand amperes. The arc appearance is charac-

terized by a multitude of cathode spots and a diffuse interelectrode plasma. Loss of
material from the cathode takes the form of ions and finite particles. The ion
current leaving the electrode has a value of about 10% of the arc current with mean
ion energies (in eV) exceeding the interelectrode voltage. It is the high energy
ions exiting from the electrode at high speeds (of the order of 10^5 msec^{-1}) that are
responsible for the formation of well-bonded coatings.

Poorly coated substrates show evidence of arc strike-back in the form of small pits.
The finite particles ejected from the cathode spots often result in the presence of
'macro particles' on the coated surface. The coated components produced hence
frequently yield surface finishes poorer than those of the uncoated components. In
instances where coated parts do not mate, as for example in cutting tools, surface
finish deterioration accompanying arc coating is not a significant problem. In cases
where precision components have to be protected against wear, as in the case of ball
and roller bearings, further work is necessary to improve the surface finish obtained with arc coating.

High current and low voltage discharges may also be obtained with Hollow Cathode
devices. By heating a target with hollow cathode discharge and evaporating the
target material, large coating fluxes may be obtained. By requiring the coating flux
to pass through the low voltage discharge, well ionized coating fluxes are obtained.
Since the ionization cross section of metallic vapors is large at low voltages, with
suitably configured hollow cathode discharges, well-bonded coatings are readily
produced especially when the substrate to be coated is biased. Coating systems based
on hollow cathode discharge have been shown to be useful for low temperature coating
of substrates with hard material compounds (23).

Electrons required to heat and evaporate a target material can also be obtained from
hot hollow cathode devices operated with high currents at low voltages. In the Hot
Hollow Cathode discharge coating system (24), a grounded filament within a specially
designed cathode structure is heated and the electrons extracted by applying a positive potential to a collinear anode. A coaxial magnetic field is employed to extract
the heating beam through a narrow opening communicating between the cathode chamber
and the evaporation chamber. A low voltage arc is then ignited between the filament
and an insulated portion of the hollow cathode to produce a large current flow
through the hot cathode.

In the hot hollow cathode discharge system, current densities of several kW per cm^2
are obtained at anode surfaces to efficiently melt and evaporate the anode material.
The vapor flux is effectively ionized by the low voltage electron discharge travers-
ing across the evaporation chamber. Well-bonded coatings are readily produced. Use
of reactive gases within the coating chamber enables the synthesis and deposition of
hard metal compounds (24).

A Cathodic Arc Plasma deposition system may also be used to deposit thin films for
tribological protection. In such a system, the CAP source is the cathode of a diode
discharge operating in the arc regime (25). The arc is not sustained by the background gas in the chamber but by a metal plasma generated from the source. In the
CAP-based coating system, an arc is ignited on the cathode surface over which it
moves randomly. Depending on the discharge current, one or more cathode spots are
produced. Each cathode spot is a source of electrons and ions.

The high energy density of the cathodic arc is believed to yield 60 to 80% of the
ejected material in the form of ions with energies in the 40 to 80 eV range (25).
The highly ionized and energetic coating flux produced yields well-bonded coatings

on appropriately cleaned substrates. Titanium nitride coatings have been produced suitable for wear protection. CAP source is the newest of the coating processes and is apparently insensitive to background pressure during coating. Although 'macro particle' production is likely together with coating flux generation, CAP-based vacuum coating holds considerable promise for the deposition of thin films of different compositions.

Suitability of the Different Coating Techniques for Tribological Applications:
Thin film coating techniques may be used for a variety of reasons. In the present instance, the purpose of the coating is to afford tribological protection. As noted earlier, soft coats are used to reduce friction. Hard coats are used principally to obtain wear resistance. In both cases, as the film deposited lies between the sur-faces of a loaded tribological contact, the film will have to be sufficiently well-bonded to preclude removal by frictional tractions encountered during operation. In the case of heavily loaded tribo-elements, the films deposited should not also degrade the surface finish. In addition, loss of substrate mechanical properties during film deposition requiring subsequent treatment is not desirable. With these as specific criteria, the suitability of different coating processes for tribological applications may be summarized as follows.

Pyrolytic decomposition processes, especially those for the deposition of low shear strength noble metals, is clearly well suited for soft coating. Very thin films are readily deposited at low temperatures. Since the films can be quite thin, i.e., 100 to 200 nm, to afford friction reduction, such coatings are eminently suited for tribological applications. Good surface preparation and cleaning is necessary to obtain the needed substrate-to-film adhesion strength. Gold films produced by pyrolytic decomposition are now in use in electrical contact applications. Repeated make and break cycles common in electrical contacts show that the needed adhesion strength can be obtained without difficulty.

An application involving protection against abrasive wear has been cited for pyro-lytic decomposition-based nickel plating (5). In this instance, the needed thick coating is produced slowly at low temperatures and coated foundry patterns are presently in commercial use.

Chemical vapor deposition processes now in use do not apply for soft coat deposition. A variety of CVD hard coatings are presently produced commercially and used in applications requiring exceptional wear resistance. All of them entail high coating temperatures. Hence, while CVD is an efficient and effective surface coating tech-nology for cemented carbides, the need for post-coat thermal treatment continues to be a problem for the fabrication of wear-resistant precision mechanical components made of constructional and alloy steels. Substrate mechanical properties, surface finish, and the part precision require restoration. This task is expensive, since the surfaces produced are wear-resistant.

In instances where mating tribo-elements are not present, as in tooling applications, CVD is a powerful and low cost means to obtain good tribological characteristics. Precision tribological elements needed for operation in particularly hostile operat-ing environments can justify the added cost of restoring precision and surface finish. High vacuum bearing assemblies suitable for spacecraft applications is an example. They are produced with CVD coatings and are in current use (26). Develop-ment of lower temperature CVD techniques which do not require coating temperatures in excess of approximately 550° C will most certainly lead to wide spread use of CVD for tribological applications. Low temperature techniques are likely to require refinishing solely to restore surface finish and this, in most instances, is feasible with diamond lapping.

Of the PVD techniques, DC sputtering is unsuited for tribological applications as the coating rates obtained are low. Despite this limitation, RF sputtering is in wide use for the deposition of soft coats, since thick soft coats are rarely required. Molybdenum sulfide coats produced by RF sputtering are in routine use (27). Other promising chalcogenides are presently being evaluated (21,28,29) for different tribological applications.

Ion plated metallic films have been shown to be beneficial in tribological applications (30). A wide range of titanium nitride hard coatings produced with techniques other than magnetron sputtering are often referred to as ion plated hard coats. The techniques used include ARE (17) as well as the arc coating techniques cited. In applications where mating elements are not present, the hard coats produced perform quite well. Tooling applications based on arc coating processes (22-25) have come into commercial use.

In the case of precision components, only magnetron sputtering enables non-distortive coatings satisfactory for precision applications (19-21,31). A variety of bearing applications have been demonstrated with single layer and duplex coatings. With further development of arc coating techniques, it should be possible to hard coat precision tribological components without loss of surface finish and precision.

In closing, it is noted that a major impediment in the reliable use of thin film technology in tribological applications is the absence of a quantitative substrate-to-film adhesion test. This difficulty is on the verge of resolution with a newly devised quantitative testing method (32).

Summary: In this review, the available coating technologies suitable for the tribological protection of precision engineering parts with thin films are briefly discussed, CVD, PVD, Arc Coating and other techniques are reviewed. Those suitable for tribological application are identified. The advantages and disadvantages of different techniques are briefly presented.

Materials that do not experience significant metallurgical changes at temperatures over 800° C can be satisfactorily coated with CVD, salt bath electrosynthesis and pyrolytic deposition techniques in order to obtain tribological protection. Iron- and steel-based components may be coated with one or more of the vacuum coating techniques at 'low' temperatures. A number of processes are available to protect non-mating engineering parts. Tool coating is an example. Several of the arc coating techniques and magnetron sputter coating are then satisfactory. They are in commercial use.

For precision components, only DC and RF magnetron sputtering have so far been demonstrated to provide well-bonded coatings wich do not require refinishing prior to use. Ion plating and RF sputtering are also satisfactory in such applications provided that the needed coating thic ness is below 1000 nm. This requirement is met in soft coat applications. As hard coatings require film thicknesses of 1000 to 5000 nm or more, further process development is needed to overcome some of the current limitations characteristic of these processes in order to allow their use to provide tribological protection to precision components.

References:

1. Anantha, N.G., et al., Proc. 2nd Int. Conf. on C.V.D., The Electrochemical Society, Princeton, NJ, 1970, p. 649.
2. Withers, J.C. and L.C. McCandless, ibid., p. 393.
3. Fitch, H.M., U.S. Patent No. 2,984,575, 1961.
4. Fitch, H.M., U.S. Patent No. 3,262,790, 1966.
5. Vasilash, G.S., Manufacturing Engineering, $\underline{84}$, 1980, p. 25.
6. Hough, R.L., Proc. 3rd Int. Conf. on C.V.D., Am. Nuclear Society, Hinsdale, IL, 1972, p. 232.
7. Sjostrand, M.E., Proc. 7th Int. Conf. on C.V.D., The Electrochemical Society, Princeton, NJ, 1979, p. 452.
8. Spear, K.L., ibid., p. 1.
9. Erickson, G., Acta Chem. Scand., $\underline{25}$, 1971, p. 265.
10. Teyssandier, F. and M. Ducarvoir, Proc. 7th Int. Conf. on C.V.D., The Electrochemical Society, Princeton, NJ, 1979, p. 398.
11. DeGee, A.J., et al., paper presented at the IRG O.E.C.D. Meeting, September 1982, L.S.R.H., Neu Chatel, Switzerland (to appear in $\underline{Wear}$, 1983).
12. Hintermann, H.E., et al., Wear, $\underline{48}$, 1978, p. 262.
13. Andrieux, J.L. and G. Weiss, C.R. Acad. Sci., $\underline{219}$, 1944, p. 550.
14. Koyama, K., et al., J. Electrochem. Soc., $\underline{130}$, 1983, p. 147.
15. Hockman, A.J. and R.S. Feigelson, J. Electrochem. Soc., $\underline{130}$, 1983, p. 221.
16. Stern, K.H. and S.T. Gadomski, J. Electrochem. Soc., $\underline{130}$, 1983, p. 300.
17. Raghuram, A.C. and R.F. Bunshah, J. Vac. Sci. Tech., $\underline{9(6)}$, 1972, p. 1389.
18. Hieronymi, R., et al., Thin Solid Films, $\underline{96}$, 1982, p. 241.
19. Ramalingam, S. and W.O. Winer, Thin Solid Films, $\underline{73}$, 1979, p. 267.
20. Ramalingam, S., et al., Thin Solid Films, $\underline{80}$, 1981, p. 297.
21. Ramalingam, S., et al., Thin Solid Films, $\underline{84}$, 1981.
22. Sablev, L.P., et al., U.S. Patent No. 3, 793,178, 1974.
23. Moll, E., U.S. Patent No. 4,197,175, 1980.
24. Nakamura, K., et al., Thin Solid Films, $\underline{40}$, 1977, p. 155.
25. Johansen, O.A., A.R. Lefkow, and W.M. Mularie, to appear in Thin Solid Films, 1983.
26. Gass, H. and H.E. Hintermann, Proc. 4th Int. Conf. on C.V.D., The Electrochemical Society, Princeton, NJ, 1973, p. 563.
27. Christy, R.L. and G.C. Barnet, Lubrication Engineering, $\underline{34}$, 1978, p. 437.
28. Bergmann, E., et al., Tribology International, $\underline{14}$, 1981, p. 329.
29. Montes, H., et al., Journal du Froltement Industriel, No. 16, July 1982, p. 3.
30. Todd, M.J. and R.H. Bentall, Proc. 2nd Int. Conf. on Solid Lub., ASLE, Park Ridge, IL, 1978, p. 148.
31. Rigert, R.P., AFML Technical Report No. TR-78-192, Air Force Materials Lab., WPAFB, Ohio, 1978.
32. Ting, B., M.S. Thesis, Georgia Institute of Technology, Atlanta, GA, 1983.

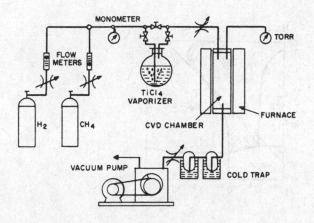

Figure 1.   Schematic Illustration of a Typical CVD System

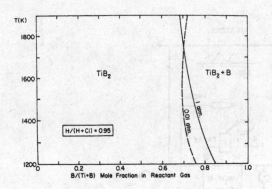

Figure 2.   The CVD Phase Diagram for Titanium Diboride Deposition

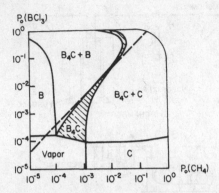

Figure 3. The CVD Phase Diagram for Boron Carbide Deposition at 1800 K and 1 Atmospheric Pressure

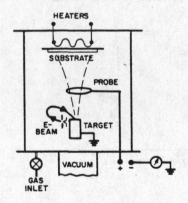

Figure 4. Schematic Illustration of a PVD System

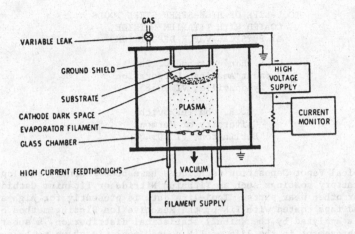

Figure 5. Schematic Illustration of an Ion Plating System

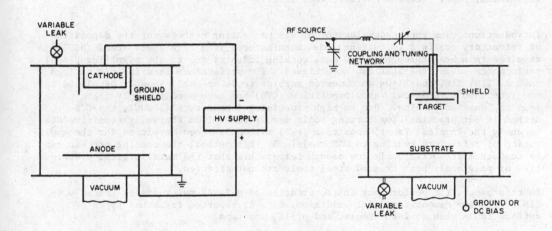

Figures 6 and 7. Schematic Illustration of DC and RF Sputtering Systems.

TOOL-LIFE OF HIGH-SPEED STEEL TOOLS
COATED WITH TITANIUM NITRIDE
BY PHYSICAL VAPOR DEPOSITION

W. E. Henderer and G. Thomas
Vermont American Corporation
Lyndonville, VT 05851

B. F. von Turkovich
University of Vermont
Burlington, VT 05401

Abstract: Physical Vapor Deposition techniques have been recently developed for
depositing refractory coatings such as Titanium Nitride or Titanium Carbide on
cutting tools or other wear parts. Tool-life data is presently for high-speed
steel drills and taps coated with TiN by the Reactive Ion Plating method of PVD.
When the data is analyzed by the Weibull statistical distribution, a substantial
and reliable improvement in the life of TiN coated tools is shown to be possible.
The response of tool-life to cutting speed and feed has been evaluated for untreated
and TiN coated HSS drills. It is demonstrated that TiN coated HSS drills can be
operated at significantly greater cutting speeds or feeds than uncoated drills.
Although the TiN compound has a number of physical properties which may explain
these results, cutting force measurements indicate that TiN coating reduces the
specific cutting energy by reducing friction at the tool/chip interface.

Key words: Coating; Cutting tools; drills; high-speed steel; physical vapor
deposition; taps; titanium nitride; tool-life.

Introduction: The recent development of vacuum coating methods for the deposition
of refractory coatings to cutting tools, manufactured from High-speed Steel (HSS), has
resulted in a breakthrough to the metalworking industry due to the significant tool
performance advantages that can be achieved. The performance benefits of coatings
such as TiN, TiC, and Al_2O_3 on cemented carbide tools are well known. Developed in
the late 1960's, Chemical Vapor Deposition (CVD) has been used to apply these coatings to cemented carbide. Due to high processing temperature (1000C), the CVD
method is not practical for cutting tools manufactured from thermally sensitive HSS.
Recently the Physical Vapor Deposition (PVD) method has been developed for the application of refractory coating to HSS tools. By this method, the coating process can
be conducted at sufficiently low enough temperature that the metallurgical properties of previously heat treated steel tools are not affected.

In this paper, the performance characteristics of M-7 HSS twist drills coated with
TiN by PVD are demonstrated. In addition, data is reported for other types of
cutting tools such as lathe tools, end mills, and taps.

Analysis: A variety of PVD methods have been developed for reactive deposition under
vacuum. They are primarily distinquished by the method of creating a metal vapor:
sputtering, arc discharge, or evaporation. The utilization of a gas plasma (or glow
discharge) is common to all PVD techniques. If a negative electrical bias is applied
to the substrates prior to and during deposition, the method is known as ion plating.

18

Excellent adherence of the coating to the substrate is a characteristic of the Reactive Ion Plating (RIP) method. For example, reliability analysis has been made using the Weibull statistical distribution for the tool-life of HSS twist drills coated with TiN by RIP, Figures 1 and 2. In this case, the tool-life was determined for samples of approximately five drills each chosen randomly from five coating charges and two manufacturing lots. Not only was the median tool-life (B50) substantially increased by TiN coating, but in addition the Weibull slope (b) of the sample of coated drills was equivalent to Weibull slope of the sample of uncoated drills. The latter observation indicates that extremely reliable performance can be achieved by Reactive Ion Plating, and the tool-life variability that is obtained is due to variation in drill manufacture or non-uniformity of the work material.

The potential for increased productivity is the primary economic advantage of TiN coated HSS tools. The response of tool-life to cutting speed and feed has been determined for 1/4 inch (6.35mm) diameter M-7 HSS drills coated with 1.5 micron thick TiN.[1]   Figure 3 shows the response of tool-life of uncoated and TiN coated HSS drills to speed over the range of 44 to 193 SFM (13 to 59 m/min.) at a constant feed rate of 0.0045 ipr (0.11 mm/rev.) in AISI 4340, 32HR$_C$. As evident in Figure 3, TiN coating substantially increases tool-life. The factor of improvement depends on speed and increases with increasing cutting speed. Considering that the normally recommended cutting speed for uncoated drills is 50 SFM (15 m/min.) under these conditions, an equivalent tool-life of TiN coated drills is reached at 100 to 125 SFM (30 to 38 m/min.) Figure 4 indicates that the feed can be similarly increased with TiN coated drills. In AISI 4340 (32HR$_C$), TiN coated HSS drills can be operated at 93 SFM (28 m/min.), 0.0078 ipr (0.20 mm/rev.) and obtain tool-life equivalent to uncoated drills when operated at 44 SFM (13 m/min.), 0.0031 ipr (0.08 mm/rev.).

Surprisingly, it should be observed in Figure 4 that the tool-life of TiN coated HSS drills increases over the feed range 0.002 to 0.008 ipr (0.05 to 0.20 mm/rev.) rather than substantially decrease as with uncoated drills. Konig[2] conducted similar tests of HSS drills coated with TiN by Physical Vapor Deposition in 42 Cr-Mo-4V steel (AISI 4137). Those results are directly comparable to the data presented here including the response of tool-life to feed for TiN coated drills.

It is well known that tool wear is highly sensitive to cutting temperature. The temperature at the chip-tool interface is dependent on the specific cutting energy, the cutting speed, and feed rate. In these tests, abrupt tool failure occurs when the cutting edge corner collapses. This is related to the loss of high temperature strength of the cutting edge due to time and temperature dependent tempering of the HSS structure.

In order to determine whether the specific cutting energy is affected by the TiN coating of HSS drills, cutting force measurements have been made over the same speed and feed range as the tool-life test, Figures 5 and 6.[1] Figure 5 shows that drill torque and thrust is relatively constant over the range in cutting speed investigated. The torque and thrust of TiN coated HSS drills was 35 and 43 percent lower, respectively, than uncoated drills. Figure 6 shows the response of drill cutting forces to increase approximately linearly with feed. Torque and thrust was, however, less sensitive to feed than uncoated drills.

The reduction in torque of TiN coated HSS drills implies a similar reduction in specific cutting energy and therefore substantial reduction in cutting temperature. The lower sensitivity of torque and thrust implies a substantial reduction in friction. Formation of built-up edge (BUE) or adherence of work material to the cutting edge is a prevalent occurrence when cutting metals. However, SEM inspection of chips

19

produced during drilling shows that BUE is nonexistant on TiN coated drills when cutting steel.[1] This observation confirms the substantial reduction in friction at the chip/tool interface with TiN coating. This reduction in adhesion between the work material and the TiN interface can be attributed to the insolubility of the TiN compound in steel.

Compounds such as TiN obviously have other notable characteristics which inhibit wear. Adhesive wear would obviously be reduced in this case. Since TiN is a refractory compound, wear by abrasion would be reduced by the resistance of the TiN interface to localized deformation at high temperature. It has also been suggested that since tool wear is a thermally activated process, the chemical stability of the tool surface is a controlling factor. As indicated by the free energy of formation, TiN is a chemically stable compound and therefore wear by solution could be reduced.[3]

Substantial performance benefits have also been demonstrated for other types of High-Speed Steel cutting tools coated with TiN by Physical Vapor Deposition. Since thread generating taps are a sizing tool, thread tolerances can be maintained for longer periods of time since TiN coating not only resists wear which produces under-size gaging; but in addition, the coating resists galling of work material to tap thread flanks and therefore reduces the tendency for oversize gaging. Performance benefits have been established for TiN coated HSS hobs, broaches, reamers, and end mills as well.[4]

Summary: Development of the Physical Vapor Deposition technique for low temperature coating of HSS tools with refractory compounds, such as TiN, has resulted in significant and reliable performance benefits for the metalworking industry. The significant economic advantage in machining is the capability to substantially increase cutting speeds and feeds and therefore improve machine tool productivity. The data given here for TiN coated HSS drills demonstrates this potential. Although TiN coating has a number of properties which reduce wear mechanisms, this data suggests that the primary effect is the reduction of specific cutting energy. This results from the reduction of adhesion of work material to the TiN interface due to the insolubility of TiN in steel.

References:

[1] W. E. Henderer, "Performance of Titanium Nitride Coated High-Speed Steel Drills," Proc. NAMRC XI, May, 1983, 337-341.

[2] W. Konig, D. Lung, Th. Wand, "fur Leistungsfahigkeit TiN-Beschichteter HSS - Spiralbohrer," Industrie (W. Germany), Vol. 27 (1982), 10-14.

[3] B. M. Kramer, N. P. Suh, "Tool Wear by Solution: A Quantitative Understanding," Trans. A.S.M.E., J. Engr. Industry, Vol. 102 (1980), 303-309.

[4] R. L. Hatschek, "Coatings: Revolution in HSS Tools," American Machinist, Special Report 752, Vol. 127, No. 3, March 1983, 129-144.

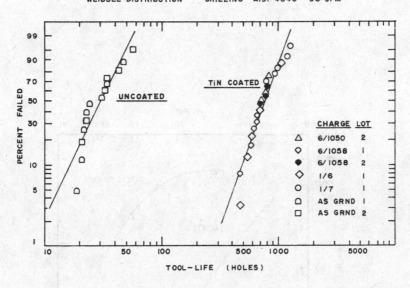

Figure 1. Weibull Tool-Life Distribution, 93 SFM

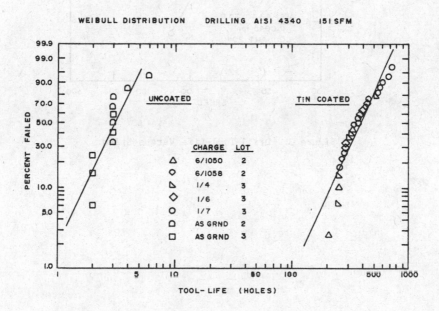

Figure 2. Weibull Tool-Life Distribution, 151 SFM

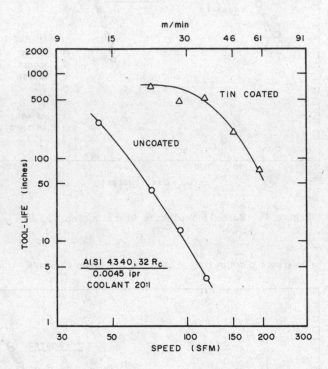

Figure 3. Drill Tool-Life Versus Speed

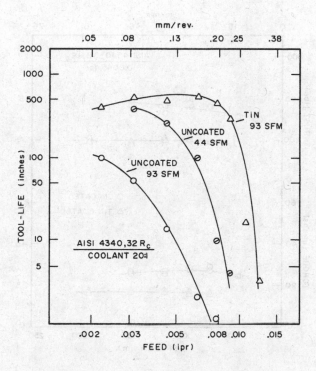

Figure 4. Drill Tool-Life Versus Feed

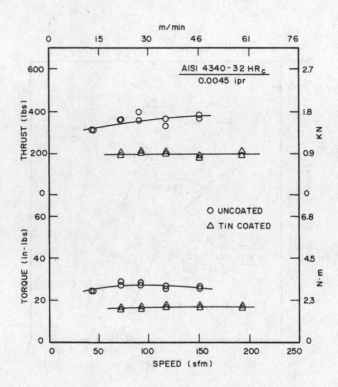

Figure 5. Drill Torque and Thrust Versus Speed

24

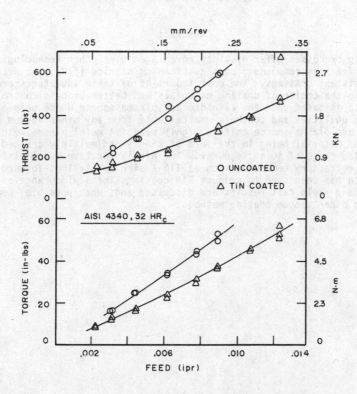

Figure 6. Drill Torque and Thrust Versus Feed

ARC COATING PROCESSES

A. Lefkow and W. M. Mularie
Vac-Tec Coating Services
St. Paul, Minnesota 55117

Abstract

Vacuum coating processes offer distinct advantages over other technologies used in depositing refractory compounds, such as titanium nitride (TiN), for wear and corrosion resistant coatings. The most important of these advantages are more flexibility in the choice of coating materials and better process control. One such vacuum process is based upon the cathodic arc plasma source whose unique properties offer coating qualities and controls unattainable from any other vacuum deposition source. The arc plasma source emits a highly reactive metal plasma with up to 80% of the emitted material being in the form of singly and multiply charged ions with energies in the range of 40 ev to 80 ev. The highly reactive metal plasma makes synthesis of refractory compounds such as TiN a simple, straight-forward process. The arc source has been used to produce TiN coatings for a wide range of applications. Experimental and field test results are discussed and, when possible, compared to results using other vacuum coating methods.

STATE OF THE ART IN COMPOSITE ELECTROLESS COATING

N. Feldstein and T. Lancsek
Surface Technology, Inc.
Princeton, New Jersey 08540

ABSTRACT: The ability to codeposit particulate matter in electrolessly deposited matrices has led to a new generation of composite coatings. These offer a challenge to the plater and designer to incorporate deposits having unique physical and/or chemical properties.

Particulate matter that can be codeposited ranges from dielectrics and metals to compounds and metal alloys and combinations of these. Properties of the composite coatings are particularly dependent upon the concentration, particle size, particle size distribution, and the nature and morphology of the particulate matter, as well as the electroless plating bath employed.

From the great variety of particulate matter that can be codeposited, attention has thus far been focused primarily upon aluminum oxide, polycrystalline diamond, silicon carbide, and polytetrafluoroethylene (PTFE). This paper reviews the state of the art and its evolution, including results depicting the merits of these coatings in comparison to conventional coatings.

INTRODUCTION: The ability to codeposit fine particulate matter within an electroless metal deposit has led to a new generation of composite coatings.

By contrast to electrolytic plating, electroless metal deposition is a chemical method of deposition which does not require the use of auxiliary power supplies, anodes or cathodes. Deposits derived from the electroless deposition technique are generally limited to a few metals (e.g., nickel, cobalt, copper, gold, and silver) and a variety of alloys which possess unusual physical and/or chemical properties. It is these unusual properties and the simplicity of the plating procedure that have led to industry's acceptance of this deposition technique.

Successful codeposition is dependent on various factors, including particle catalytic inertness, particle charge, electroless bath composition, compatibility of the particles with the metallic matrix, plating rate, and particle size distribution.

The mechanics of composite electroless plating are contrary to practices prevailing for conventional electroless plating. Foreign, finely divided, solid particulate matter is added to and dispersed throughout the electroless plating bath, even though the plating bath is thermodynamically unstable and is prone to homogeneous decomposition. The dispersed particles are not filtered out. The dispersion of the particulate matter results in a new surface area loading in the range of $100,000 \, cm^2/L$, which is some 800X greater than the plating load generally acceptable in conventional electroless nickel plating.

27

Though composite electroless plating is still in its infancy, future prospects are most attractive to the user as well as to the designer. The adaptation of a coating which offers improved wear and/or lubricity can yield ways by which both energy and natural resources can be conserved. The use of composite electroless coating also offers the benefit of reduced solution handling and pollution problems, as well as reduced reliance on strategic materials (e.g., chromium) for wear applications.

HISTORICAL DEVELOPMENT: Some types of composite plating were demonstrated in earlier work:

Composite electrocoating (CEM) has been commercially practiced for many years[1] to obtain coatings for abrasive and wear-resistance applications. In CEM finely divided particles are dispersed along with the electrolytic plating composition; the codeposition takes place with the application of a cathodic potential.

In pursuing improved corrosion resistance for nickel-chromium electrodeposits, Odekerken[2] interposed the structure with an intermediate layer containing finely divided particles distributed within a metallic matrix. Electroless nickel deposition was demonstrated as a means of depositing the intermediate layer, utilizing finely divided aluminum oxide, polyvinyl chloride resin (PVC), as well as binary mixtures of powders, e.g., PVC along with aluminum oxide. Though Odekerken employed a relatively thin layer for the intermediate layer, thicker layers could be attained with longer immersion periods due to the autocatalytic nature of electroless plating.

Metzger et al[3] , in 1966, duplicated an electroless nickel containing alumina (Al_2O_3) particles. Their efforts were then directed towards the codeposition of silicon carbide, with the first commercial application of the coating on the Wankel (rotary) internal combustion engine. The Metzger et al efforts are documented in a German patent application[4] which, though it did not issue as a German patent due to public opposition, was later modified in the U.S., and led to several patents.

In typical composite coatings, the fine particulate matter can be selected in the size range of 1μ to 10μ, with a loading of up to 50% by volume within the matrix. The ratio of codeposited particles to the metal matrix in composite electroless plating can be adjusted to a fixed and constant ratio. Most commercial practices, however, appear to focus on 25% to 30% by volume of the particle within the matrix. Due to the multilayer manner of codeposition these coatings are the "regenerative type." Thus the coating still maintains its properties despite the removal of one or more of the layers.

Coatings of 0.5 to 1 mil in thickness have been found adequate for most commercial applications though it is possible to generate a thicker coating.

Though it would appear that a wide variety of particulate matter can potentially be codeposited, at present commercial composite electroless plating activities have been limited to just a few types of particulate matter: diamond, silicon carbide, aluminum oxide, PTFE. Even within these types of particles only certain specific forms have been found of utility: the polycrystalline form of diamond, and the platelet form of alumina.

Currently there are at least two commercial coatings of electroless nickel which contain silicon-carbide as the finely divided particulate matter[5,6]. In 1981 a commercial composite coating bearing PTFE was introduced[7]. Today, however, diamond appears to be the most popular particle for use in composite electroless plating, a trend that evolved as a result of the discovery and availability of a new form of diamond, i.e., a polycrystalline type produced by a shock synthesis technique.

Early attempts with man-made diamond were not fruitful. Failure and homogeneous decomposition of the electroless plating baths resulted. Though it has not been confirmed, it is our belief that the early difficulties reported with diamond may have been due to the composition of the diamond, stemming from its manufacturing procedures. Man-made diamond incorporates catalytic metals such as nickel, cobalt, copper, and iron, as well as mixtures of these during the synthesis process. Their presence in the electroless plating bath may be responsible for the disastrous bath decomposition.

Not until the pioneering efforts of Christini et al[8] was successful deposition of a composite diamond coating achieved. Their work was accomplished through the incorporation of a new, man-made polycrystalline diamond which led to an exceptionally wear resistant coating along with excellent anchoring within the matrix.

The development of electroless composites incorporating PTFE were impeded as a direct result of the hydrophobic nature of this resin in aqueous solutions, and consequently its inability to wet and properly be dispersed in the plating bath, and probably its affinity to metallic matrices.

By contrast to the difficulties encountered with diamond and PTFE particles, composites incorporating aluminum oxide or silicon carbide were achieved without serious difficulty.

The first technology center for composite electroless plating was formed in 1981[9]. Its purpose is to provide information and guidance for potential applications, measurements, specifications and the individual handling requirements for each form of particulate matter, since an electroless bath suitable for one specific particulate matter may require modification(s) when it is to be used for other particulate matter.

Figure 1 shows a photomicrograph of a cross-sectional cut of a commercial composite electroless nickel containing alumina particles having an average particle size of about 4.5μ. The preferred alumina particles are platelet shaped[10].

Figures 2 and 3 show photomicrographs of a cross-sectional cut of a composite coating containing polycrystalline diamond with average sizes of 1½μ and 6μ, respectively.

Figure 4 shows a photomicrograph of a cross-sectional cut of a composite electroless nickel containing silicon-carbide of a nominal size of 7 μ and a coating thickness of 1 mil[6].

Figure 5 shows a scanning electron micrograph (SEM) of a cross-sectional cut of a composite electroless nickel coating containing finely divided PTFE particles.

Figure 6 shows a cross-sectional cut of a composite coating containing two distinct particulate matter, i.e., aluminum oxide in combination with PTFE.

WEAR RESISTANCE: Thus far the primary objects of composite electroless coatings have been to improve the wear resistance and/or corrosion resistance and/or lubricity of machinery parts. Although the basic mechanism for the incorporation of the insoluble particulate matter into the metallic matrix is not fully understood, highlights of the dominating parameters have been characterized in various reports[8,11-16]

Various test procedures have been employed to evaluate the degree of wear resistance achieved: the Taber Wear Test; the Alfa Wear Test; and the Accelerated Yarnline Wear Test developed at E. I. du Pont de Nemours & Company.

The Taber Wear Test evaluates the resistance of surfaces to rubbing abrasion produced by the sliding rotation of two unlubricated abrading wheels against the rotating sample. Figure 7A shows a schematic of the key components in the Taber Wear Test.

In the Alfa Wear Test samples are subjected to wear against a hardened steel ring under clean and lubricated conditions. Figure 7B shows the key components in the Alfa Wear Test.

The Accelerated Yarnline Wear Test was designed to simulate typical conditions commonly found in textile machinery. Figure 7C shows the key components in the Accelerated Yarnline Wear Test.

In all of these laboratory tests it is generally the scar depth or the worn volume that is measured. It is particularly important in compiling comparative data from these tests that all test specimens be of similar morphology. Variations in particle size, concentration of loading, and surface finish could yield misleading and erroneous wear test results. Surface finish of the coating is especially important in the Taber Test.

TABLE I

ACCELERATED YARNLINE WEAR TEST RESULTS (8)		
Test No. Material	Test Time min.	Wear Rate μ/hr
1 Electroless Ni-B/9-μ Diamond "A" Composite Coating (polycrystalline)	85	5.1
2 Electroless Ni-B/9-μ Natural Diamond Composite Coating	85	10.2
3 Electroless Ni-B/9-μ Diamond "B" Composite Coating	85	13.1
4 Electroless Ni-B/8-μ Al₂O₃ Composite Coating	9	109
5 Electroless Ni-B/~ 10μ SiC Composite Coating	5	278
6 Electroless Ni-B As-Plated (with no particles)	1/30	23,000

Table 1 provides wear test results for various composite electroless coatings based on the Accelerated Yarnline Wear Test. In the last column wear rate (μ/hr) is provided; the higher the number, the greater the deterioration in the wear resistance.

Three different types of diamond particles were tested since they were all commercially available at the time of the Christini et al investigation [8]. The "A" diamond of Test 1 was a polycrystalline diamond prepared in accordance to Cowan [17]. Test 2 used natural diamond. The "B" diamond of Test 3 was a General Electric synthetic diamond believed to have been synthesized in accordance with U.S. Patents 2,947,608 through U.S. 2,947,611. Tests 4 and 5 employed composite coatings bearing aluminum oxide and silicon carbide, respectively, prepared in an electroless nickel-boron bath. Test 6, used as a basis for comparison, is an electroless nickel-boron coating, the same matrix composition as used in Tests 1 through 5, but with no particulate matter codeposited.

The wear rates were determined after a specific time interval, as seen in column 3. The results demonstrate that diamond "A" is superior to all the other diamond particulate matter evaluated, as well as being the best of all the composite coatings regardless of the particulate matter occluded.

Comparison of the results obtained for the aluminum oxide composite and the silicon carbide composite shows a trend which appears to be contrary to previous reports [14] and, perhaps, to reasonable anticipation. The inclusion of the aluminum oxide appears to provide superior wear even though the testing time was longer (9 minutes as compared to 5 minutes) and the particle size was smaller (8μ as compared to 10μ). Greater wear resistance is generally obtained with larger particulate matter. It should be remembered that the hardness of silicon carbide is about double the hardness of aluminum oxide. This behavior, then, though not fully understood, may be attributable to the manner in which the particles are anchored within the metallic matrix, i.e., the compatibility of the particles within the matrix, as well as their resistance to being pulled out of the matrix during the wear mode [8]. It is noted that all the composites, regardless of the particulate matter incorporated, performed substantially better than the electroless nickel (Test 6) without any particulate matter. The merits of composite electroless coating in comparison to conventional electroless coating are well demonstrated by Parker [18,19] using various wear testing techniques. Parker measured the wear resistance of miscellaneous composites bearing carbides, borides, diamond, aluminum oxide, and Teflon. It appears, however, that the composites he tested employed particulate matter of different sizes, and neither the concentration of particles within the matrix nor the surface roughness for the coating prior to testing were revealed. Accordingly, no definitive conclusions can be made as to the preference of one particle over another based upon the data he published [18,19].

The difference in the results between the three types of diamond is unexpectedly revealing. Further investigation of this phenomenon (Table II) provided insight, and a model was postulated to account for the outstanding behavior of the polycrystalline diamond in comparison to the other diamonds as well as to the other composites.

31

TABLE II

Test No.	DIAMOND PULL-OUT FROM ACCELERATED YARNLINE WEAR TEST GROOVES Ni-B Alloy Composites*** (8)	Area I*	Area II**	Percent Difference	Comments
1	Diamond Count for Diamond "B"	50	34	-32%	⎡ Difference primarily due to diamond re- moval from the bottom of the wear grooves.
2	Diamond Count for Natural Diamond	60	45	-25%	⎣
3	Diamond Count for Diamond "A" (Polycrystalline)	42	46	+9.5%	Difference due to the randomness and un- covering of diamonds on the bottom of the wear groove which were just under the surface of Ni-B alloy matrix.

   * Side of wear groove and adjacent as-plated composite surface.
  ** Bottom of wear groove, area where the majority of wear occurs.
*** Prepared same as in Table I, Tests 1, 2, and 3.

Table II provides measurements, made on a quantitative basis using scanning electron microscopy technique, for the particle pull-out magnitude of the three types of diamonds evaluated. The areas examined were along the wear groove, the adjacent as-plated composite surface, as well as the bottom of the wear groove where the majority of the wear occurred.

A very considerable decrease in the diamond count was observed both for diamond "B" (Test 1) as well as for the natural diamond (Test 2). Yet there was *no* decrease in the diamond count in the Test 3 results of the polycrystalline "A" type diamond in the critical areas examined.

These Table II results reinforce the conclusions drawn from Table I, suggesting that the wear resistance for composite coatings are de- pendent not only on the *nature* of the particle used and its ability to be wetted by the matrix, but also on the compatibility and/or inter- action of the particle with the matrix which relates to the particle's morphology. This latter compatibility may lead to an "easy" or a "hard" pull-out characteristic for the particles from the matrix. Consequently, based on the Table I and Table II data, it has been con- cluded that the "A" type polycrystalline diamond is much more firmly secured within the metallic matrix than are other diamond particles or other particulate matter (silicon carbide and aluminum oxide); this accounts for its outstanding wear resistance as noted.

In the tests of Table I and Table II the matrix used was electroless nickel-boron. However, other electroless plating compositions can be used resulting in different alloys and/or metals (e.g., copper, cobalt, nickel-phosphorous, nickel-cobalt-boron, etc.).

Table III shows the results of changing the metallic matrix of Table I to a Ni-P type alloy. A decreased wear resistance is noted as a result of the matrix change.

TABLE III

Test No.	Composite Coating		Test Time Min.	Wear Rate Hr.
	WEAR RATE IN A NICKEL-PHOSPHOROUS MATRIX (8)			
1	Electroless Ni-P/1μ	Diamond "A"	2	378
2	Electroless Ni-P/1μ	Natural Diamond	2	732

Despite the decreased wear resistance due to the change of matrix
documented in Table III, as compared to Table I (Tests 1, 2, and 3)
and Table IV (Test 6), nevertheless the polycrystalline diamond still
reveals its superior behavior independent of the metallic matrix in
which it is codeposited.

TABLE IV

Test No.	WEAR RESISTANCE vs. PARTICLE SIZE AND PERCENT LOADING OF PARTICLES (8)			Wear test data
	Dispersed phase data			
	Average Particle Size,μ	Volume,%	Time,Min.	Rate, μ/hr.
1	12-22*	20	85	3.4
2	9	20	85	5.1
3	5	20	85	6.2
4	3	29	30	11.6
5	3	5	10	65
6	1	20	2	216

* Particles were selected and tested in the range of 12-22 micron.
Note: Ni-B matrix was used with the codeposited polycrystalline
diamond. Testing was via the Accelerated Yarnline Wear Test.

From the results of Table IV the following trends are deduced: 1. Wear
resistance is related to particle size. However, increase in particle
size will yield an optimum point in the obtainable wear resistance
values; beyond the 10μ particle size there does not appear to be any
discernible gain for wear resistance. 2. Wear resistance is
related to volume percent (loading) of the codeposited particulate
matter in the electroless bath.

Field results comparing relative wear between an electroplated
nickel bath containing 20% by volume of polycrystalline diamond with
an electroless nickel-phosphorous bath also containing 20% by volume
of polycrystalline diamond show that the latter results in increased
wear resistance of about 40%, with further increase possible by
heat treatment of the electroless composite.

TABLE V

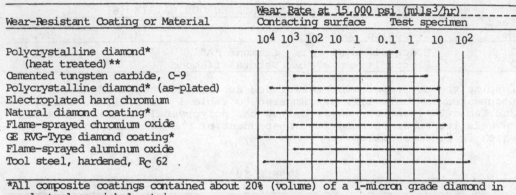

Wear-Resistant Coating or Material	Wear Rate at 15,000 psi (mils³/hr) Contacting surface	Test specimen

| | 10⁴ 10³ 10² 10 1   0.1 1   10 10² |

Polycrystalline diamond*
  (heat treated)**
Cemented tungsten carbide, C-9
Polycrystalline diamond* (as-plated)
Electroplated hard chromium
Natural diamond coating*
Flame-sprayed chromium oxide
GE RVG-Type diamond coating*
Flame-sprayed aluminum oxide
Tool steel, hardened, R$_C$ 62

*All composite coatings contained about 20% (volume) of a 1-micron grade diamond in an electroless nickel matrix.
**750°F for 90 min. in air.

In Table V some typical results of the Alfa Wear Testing are noted for both the test specimen (coated sample) and the contacting surface. Once again, the superiority of the polycrystalline diamond is noted. In fact, the relative wear for the other diamond composites (natural and GE RVG-type) appear much higher than the previous testing results.

In Table VI typical results from the Taber testing are shown. It is interesting to note that, based on the results of Table VI, the wear lifetime for the composite diamond coating utilizing the polycrystalline diamond is expected to be four times better in comparison to hard chromium plating. Field testing has more than corroborated this relative trend.

TABLE VI

	Wear Rate	
	Per 1,000 cycles	Relative to
Wear Resistant Coating or Material	(10^4mils³)	diamond
Polycrystalline diamond*	1.159	1.00
Cemented tungsten carbide Grade C-9 (88 WC, 12 Co)	2.746	2.37
Electroplated hard chromium	4.699	4.05
Tool steel, hardened, R$_C$ 62	12.815	13.25

*Composite coating contained 20 to 30% of a 3-micron grade diamond in an electroless nickel matrix.

SURFACE FINISH: Inclusion of particulate matter within the electroless metal matrices increases the surface roughness. The degree of surface roughness is dependent upon various parameters such as particle size, concentration of loading, thickness of coating, particle size distribution, and smoothness of the substrate. Since it is important to take the ultimate application of the plated article into account, surface roughness must be a consideration. In certain circumstances a surface finish operation may be required to modify the as-plated roughness, thereby controlling the frictional properties of the coating as well.

Table VII provides the surface roughness of composite coatings containing polycrystalline diamond of different nominal sizes in a nickel-phosphorus matrix, as well as some final roughnesses attainable with various newly developed smoothing techniques[20,21].

TABLE VII

SURFACE ROUGHNESS vs. DIAMOND PARTICULATE SIZE		
Diamond Size (Microns)	Roughness Values* (Microinches)	
	Initial (As-Plated)	Final (After Smoothing)
6	40 – 45	15 – 18
4	26	13.5
3	19	12
1½	7.2	–

*Substrates used had 1 to 2 microinches in roughness

Composite coatings, especially those containing wear particles, are difficult to smooth. The newly developed methods afford a simplicity and economy of the smoothing operation not previously available. One approach to the attainment of smoother coatings has been demonstrated by Feldstein[19]. The newly developed method relies upon the deposition of a second metallic layer which covers all exposed particulate matter and then removal of a portion of the secondary layer while still covering the exposed particles. In so doing, a smoothness level can be attained in a relatively shorter period of time (e.g., 30 seconds vs. 480 seconds).

Figure 8 A shows a scanning electron photomicrograph of a composite diamond coating as-plated (surface roughness of 35AA). For comparison, Figure 8 B shows the surface after smoothing (surface roughness of 16 AA).

It is the additional smoothing procedures that have made certain of the composite coatings adaptable to applications requiring precise frictional levels.

APPLICATIONS: To date, most successful commercial applications have primarily utilized the polycrystalline diamond or silicon carbide. The utilization of alumina or PTFE is too new to permit accurate predictions of future commercial acceptance. The following are just a few of the applications utilizing composite coatings in which great success has already been achieved:

A wide variety of machinery parts in the Textile industry; mechanical shaft seals for automotive water pumps and chemical process pumps; tools for processing glass or ceramic-filled resins; woodworking tools, such as taps, dies, drills, routers, rasps and files; molds;

and high-speed paper-handling equipment like sorters, lifters and
guides. Other uses include magnetic tape recording heads and guides
for computers; ultrasonic transducer wearplates; gage blocks; extruder
kneading blocks; screws and barrels for processing glass-filled
resins; and industrial scissors and barber scissors.

Though, as demonstrated above, the polycrystalline diamond provides a
better wear characteristic in comparison to diamond, alumina, or
silicon carbide, the temperature of operation must be considered for
each application as well. It has been demonstrated[15] via accelerated wear tests at different temperatures that the nickel-diamond
system yields a lower and relatively constant wear rate up to 300°C
in comparison to the nickel-silicon carbide in an electroless nickel
matrix. However, above 300°C, the point at which the matrix(Ni-P)
starts to soften, the advantage of the nickel-diamond disappears.
More recently, Kenton[22] has reported that composites with silicon
carbide also deteriorate in operational use above 800°F.

SUMMARY AND CONCLUSIONS: Though composite electroless plating has
been available for over a decade, its application has so far been
limited to a few particulate matter which have been applied singly,
maintaining the chemical purity of the particulate matter. It is
possible, however, to codeposit more than one chemically distinct
particulate matter simultaneously[1]. Figure 8 shows a scanning
electron photomicrograph of a cross sectional cut of a composite
electroless nickel containing PTFE and aluminum oxide. Figure 8
should be compared to Figures 5 and 1 in which these particles were
deposited alone. Some limited synergism has been noted in such binary
combinations.

Successful implementation of composite electroless plating requires
full system consideration, including the anticipated application.
The chemical nature, morphology, particle size and its distribution
should be considered along with the electroless plating composition to
be used. Though specific particles (e.g., diamond or alumina) are
available in various forms, some forms may be preferred as yielding
superior results in comparison to the inclusion of the other forms.

Much of the present data, understanding of electroless composite
coatings, and the dominating variables are credited to several
investigators [8, 11-13, 18]. Highlights of their contributions have
been documented in this publication.

The compatibility of a specific particle within a metallic matrix
appears to be dependent not only on its chemical nature and size, but
also on the physical form. Thus, it is probably both chemical and
physical factors that account for the particle cohesion within the
metallic matrix.

Though there are several laboratory procedures for the evaluation of
"wear," the "use" test is most conclusive in determining the best
coating for a specific application.

Because the techniques for composite electroless plating differ from
the techniques of conventional electroless plating, consideration must

be given to the equipment to be used. Some such efforts have already been documented [23,24].

It is anticipated that as more information concerning the outstanding properties of composite coatings becomes available, applications will proliferate and use will expand at a rapid pace.

References:
1. American Machinist, July 28 (1958) p. 80.
2. Odekerken, British Patent #1,041,753 (1966), U.S. Patent #3,644,183, and Patent DDR 41,406.
3. Metzger et al, Transaction of the Institute of Metal Finishing, 54, p. 173 (1976).
4. Metzger et al, German application B-90776 filed January 18, 1967.
5. Nye-Carbe, Registered trademark of Electrocoatings Inc.
6. Durabide, Trademark applied for by Dura-Tech Processes, Inc.
7. Nye-Lube, Trademark applied for by Surface Technology, Inc.
8. Christini et al, U.S. Patents #3,936,577 and Reissue 29,285.
9. Plating and Surface Finishing, November (1981).
10. British Patent Application #8025563 filed August 6, 1980.
11. 'Properties of Electrodeposits - Their Measurement and Significance,' The Electrochemical Society (1974), Chapter 16 by Graham and T. W. Gibbs.
12. Sharp, "Properties and Applications of Composite Diamond Coatings", Lecture presented at 8th Plansee Seminar in Reatte/Tirol, May 30, 1974; also in Proceeding: Diamond - Partner in Productivity, p. 121 (1974).
13. Berger, Machine Design, Nov. (1977).
14. Hubbell, Plating and Surface Finishing, December, 58, (1978).
15. Wapler et al, Proceeding of Diamond Conference sponsored by DeBeers, May 22-23 (1979).
16. Feldstein et al, Products Finishing, July, p. 65 (1980); Feldstein, Materials Engineering, July (1981).
17. Cowan, U.S. Patent #3,401,019.
18. Parker, Proceedings "InterFinish 1972" Basel, Switzerland.
19. Parker, Plating, September (1974).
20. Spencer, Private Communications.
21. Feldstein, U.S. Patents #4,358,922 and #4,358,923.
22. Kenton, paper presented at the ASM Surtech and Surface Coating Exposition, Dearborn, MI, May (1982).
23. Christini et al, U.S. Patents #3,853,094 and #3,940,512.
24. Kedward et al, U.S. Patent #4,305,792.

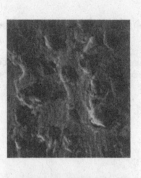

Fig. 7-A: Taber Wear Test   Fig. 7-B: Alfa Wear Test

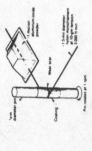

Fig. 7-C: Accelerated Yamline Wear Test

Fig. 8: Scanning electron micrograph
of composite diamond, 2000X

Fig. 8-A   Fig. 8-B
As Plated (35 AA)   After Smoothing (16 AA)

Fig. 1: Photomicrograph
of composite alumina,
4.5 micron 1200X

Fig. 4: Photomicrograph
of composite silicon
carbide, 7 micron particles,
1 mil thickness

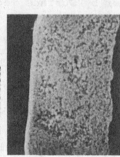

Fig. 2: Photomicrograph
of composite diamond,
1.5 micron 1200X

Fig. 5: Scanning electron
micrograph of composite
coating containing Teflon
2000X

Fig. 3: Photomicrograph
of composite diamond,
6 micron 1200X

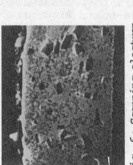

Fig. 6: Scanning electron
micrograph of composite
containing ceramic and
Teflon, 2000X

38

ELECTRODEPOSITION OF NICKEL-CHROMIUM ALLOYS

D. S. Lashmore* and Ilan Weisshaus**
Center for Materials Science
National Bureau of Standards
Washington, DC 20234

*Leader, Electrodeposition Group
**Guest worker in the Electrodeposition Group with Iscar Blades,
  Nahariyah, Israel.

Abstract: A process has been developed to electrodeposit nickel chromium alloys
from aqueous solutions. The composition of this coating can be varied from 1%
to about 60% (wt.) chromium by varying the deposition parameters. Coatings
up to 100 μm thick have been made. Dry sliding wear performance of the
20% chromium alloys is shown to be superior to the performance of electrodeposited
sulfamate nickel. The corrosion performance was characterized by the Potentiodynamic method. The alloy is a composition modulated material with layers
rich in chromium adjacent to layers poor in chromium. The layer spacings
vary from between 100 and 1000 nm. The existance of layers is consistent
with diffusion phenomena occurring in a two component system. These layers are
thought to play a role in the corrosion performance of the coating.

Key words: Alloy coatings; corrosion resistant coatings; nickel-chromium alloys;
pulse plating; stainless steel substitutes.

Introduction: The use of protective coatings is an important strategy to help
alleviate the dependence on imported sources of strategic material. The United
States imports most of its chromium and about 97% of its cobalt (1). The use
of coatings, even though they may contain a large percentage of these metals,
can result in a very large savings over bulk materials of the same average
composition (2). The coatings described in this paper are intended to serve as
possible substitutes for stainless steels.

Pulsed electrodeposition of alloys is a comparatively new technology which
allows external manipulation of the concentration of ions adjacent to the
cathode, and therefore electrochemical control over the composition of the
alloy. Pulsing also provides some control over the morphology of the coating.
Many variations of the pulse waveform are possible, the most common being a
square waveform with either a cathodic or anodic bias. Both the duty cycle,
defined as the time on divided by the period, and the frequency are typically
varied. The significance of this type of waveform lies in the fact that the
composition of the electrolyte in the depletion region adjacent to the
cathode, is strongly influenced by square glavanostatic waveforms. It is
this electrolyte composition that governs the coating composition.

Alloys containing chromium are not in common industrial use as electrodeposited
coatings, though there are many reports in the literature (3) of attempts to
produce them. The chromium-containing alloys of nickel, cobalt, and iron have

significant industrial potential because of their corrosion and wear resistance. The binary system of nickel and chromium was chosen for study in this investigation as it is the simplest chromium-containing alloy and can possibly serve as a basis for extending the process to higher order alloys. Compositions of 20 wt.% chromium are thought to represent a good compromise between corrosion resistance and reasonable ductility. It was reported by Kruger and Revez (4) that the nature of the chromium oxide on the surface of iron-chromium alloys changes from crystalline at chromium contents less than 15 wt.% to amorphous at higher chromium contents. Due to the absence of grain boundaries it has been reported that the amorphous oxide provides improved corrosion protection compared with the crystalline oxide.

Many of the processes which have been developed to electrodeposit iron-nickel-chromium alloys were discussed by Chisholm (5), who at the time concluded that there was no commercial process capable of electrodepositing the alloy. Typical of the problems that previous investigators encountered were (a) inability to achieve a thick coating, (b) inability to incorporate greater than 5% chromium into the deposit, or (c) rough, highly stressed coatings.

This paper describes a new process that has been used to pulse electrodeposit chromium-nickel alloy coatings whose chromium content were varied from 1 to over 60 wt.%, by varying the current density or the duty cycle. These coatings exhibit excellent wear and corrosion properties.

Procedure: The electrolyte used in this study made use of chloride salts of nickel and chromium. Because the chromium was present in its trivalent form the waste treatment problems characteristic of hexavalent chromium are absent. The composition of the electrolyte is:

$NiCl_2 \cdot 6H_2O$	30 g/L
$CrCl_3 \cdot 6H_2O$	100 g/L
$NaBr$	15 g/L
$Na_3C_6H_5O_7 2H_2O$ (sodium citrate)	80 g/L
H_3BO_3 (Boric acid)	40 g/L
HCOOH (formic acid)	35 mL/L
pH	3.5
T	45°C

The purpose of the sodium bromide is to keep down the content of the hexavalent chromium ions which can form as a result of oxidation at the anode. As a result some bromine is formed at the anode. The nickel is complexed by citrate, which serves to reduce the difference in deposition potential between the chromium

and the nickel. The boric acid acts to stabilize the pH and probably plays a
role in the crystal growth processes by absorption on certain crystal
planes. The formic acid acts as a buffer.

The deposition process was carried out at about 35°C and at pH 3.5 in
a pyrex cell containing about 0.5L of electrolyte, using either platinum or
high-purity carbon anodes. All of the specimens discussed here were coated
rotating disc electrodes about 1 cm. in diameter, although other diameters, as
well as other geometries, have been used without problem. The use of a rotating
disc electrode allows precise, reproducible control over the mass transport of
the electrolyte to the cathode. The wear test samples were all coated to a
thickness of 75 microns. A typical voltage current waveform is shown in Fig. 1
(photo). The current is depicted in the lower curve. During the off time the
external current is set at zero amps.

Characterization Methods: Characterization of the coating morphology, worn
surfaces and corrosion performance was done by optical metallography, scanning
electron microscopy (SEM), microhardness testing and potentiodynamic testing.
The etchant used for the optical metallography was either an aqua regia electrolytic
or 10% oxalic acid. (6V, 23°C, 5-20 sec.) Microhardness measurements were carried
out with a Knoop indenter at a load of 100 grams, using a calibration standard
periodically to insure proper operation of the instrument.

RESULTS AND DISCUSSION

I. Metallographic studies of the coatings: The as-deposited coating exhibits a
lamellar structure upon etching. This structure is shown in the optical cross
section in Fig. 2, and is a feature characteristic of many binary alloy
electrodeposits. The formation of the layered structure was discussed by
Brenner (6), and seems to result from composition fluctuations in the electrolyte
adjacent to the cathode. It should be noted that these fluctuations occur over
the interval of seconds while the pulsing frequency is on the order of microseconds.
The laminations can be observed electrochemically as shown in Fig. 3 of a
sample rotated at 200 rpm and plated at a constant current. The peaks in
voltage are seen to correspond to small voltage oscillations of about 0.1 to
1.0 mV. The frequency of oscillation approximately corresponds to the observed
layer spacing. Scanning electron micrographs of this structure, with an x-ray
line scan on the Cr.K alpha line (not shown) reveal that these oscillations are
really a result of chromium concentration gradients.

Pulsing the current affects the morphology of the coating in an important way.
These effects were discussed by Ibl (7), under conditions where metal deposition
is mass transport controlled and the thickness of the pulsating diffusion layer
is small compared to the height of the surface irregularities. These conditions
hold in the nickel-chromium system and the surface morphology illustrated in the
SEM micrographs of Fig. 4a and Fig. 4b, compares a direct current deposit
with a pulsed deposit. The duty cycle will effect the length of the pulsating
diffusion layer. As the duty cycle increases it would be expected that the
surface roughness would also increase.

II. Wear performance: The dry sliding wear testing procedure used in this
investigation was developed by W.A. Ruff (8). The coatings were formed on
uniformly heat-treated 0-2 tool steel flat blocks against which a 52100 steel
ring bearing was rotated at loads of 1 N. The apparatus was enclosed and filled

41

with argon. The coatings tested were about 75um thick. The sliding velocity
was approximately 20 cm/s, the test duration was 1 hour and the ring diameter
was about 35 cm. The instrumentation was designed to simultaneously measure
the wear rate and the coefficient of friction as a function of sliding distance
(revolutions of the bearing). Comparisons were made between electrodeposited
nickel, nickel-chromium alloy, and hardened high carbon O-2 tool steel alloy.
The lowest wear rate was found for the compositon-modulated nickel-phosphorus
coating as shown in Table I. The homogeneous nickel-phosphorus wore at more
than twice the modulated alloy rate. Nickel-chromium was comparable with the
homogeneous nickel-phosphorus. By way of comparison the wear rate for nickel-
chromium was 3 times better than that for nickel. Microhardness values up to
725 KHN, (100g) have been measured; however these wear test samples have
hardness values about 600 KHN (100g).

III. Corrosion Performance: Some accelerated corrosion tests have been made
on the pulse plated nickel chromium alloys and their performance was compared
with electrodeposited nickel and bulk 316L stainless steel sample. The results
are shown in Table II. The protection potential, defined as by the voltage value
at the intersection of the anodic and cathodic sweeps is in the neighborhood of
-100 mv SCE for nickel chromium and -130 mv for nickel. No protection potential
in this environment was found for 316L stainless. The open circuit potentials
were about -400 to -500 mv for the nickel chromium alloys depending on deposition
condition, -480 mv for nickel, -132 mv for 316L stainless. The surface after
the corrosion test for 316L stainless was severly pitted. For both nickel and
the nickel chromium alloys mixed modes of both general uniform corrosion as
well as pitting were observed. A typical potentiodynamic scan of a nickel-chromium
alloy is shown in Fig. 5.

Summary and Conclusions:

    1. A process has been developed that can produce nickel 20 wt% chromium
binary alloys.

    2. The wear performance of nickel-chromium alloys can be at least 3 times
better than nickel coatings deposited from a sulfamate electrolyte.

    3. The corrosion performance is significantly better than 316L stainless
steel in 3 wt% sodium chloride.

    4. The application of pulsed electrolysis results in greater control over
composition and surface morphology than does direct current deposition.

References

1. H. J. Grey, ASM News, vol. 10, 12 Dec. 1979 pg. 1.
2. National Materials Advisory Board; Report of the Committee on Contingency.
plans for chromium utilization; National Research Council, Nat. Ad. of Science
Washington, DC 1978.
3. P. W. Stromatt, U.S.A. patent 3,888,744 June 1975, T. Hayashi and A. Ishihama
Plating, 66 (1979) 36, M.A. Shluger and U.A. Kagakov. J. Phys. USSR 33 (1969).
4. A. G. Revz and J. Kruger., in Passivity of Metals, ed. by R. P. Frankenthal
and J. Kruger, proc. 4th International Conference on Passivity by the
Electorchem. Soc. 1978 (pg. 137).

5. C. U. Chissom and R. J. G. Carnegie, Plating and Surface Finishing, 59, 12 (1972) 28.
6. A. Brenner in Electrodeposition of Alloys. vol. 1 pg. 179. Academic press New York (1965).
7. N. Ibl, Surface Technology 10 (1980) 81.
8. A. W. Ruff and D. S. Lashmore, in Selection and use of wear tests for coatings, ASTM, STP 769, R. G. Bayer Ed. ASTM, 1982 pg. 134-156.

DRY SLIDING WEAR RESULTS (10 N load, argon, 20 m/s)

ALLOY	FRICTION COEFFICIENT	WEAR RATE (10^{-4} mm^3/m)
nickel (from a sulfamate electrolyte)	0.97	3.4
nickel-chromium 20 wt.%)	0.98	2.3
0-2 tool steel (hardened 670 KNH)	0.71	1.1
Ni-P (12 wt% P) homogeneous	0.8	1.3 ± 0.2
Ni-P (12 wt% P) Composition modulated or layered.	0.78	0.99 ± 0.16

Table I: A comparison between wear rates of nickel, nickel chromium, nickel phosphorus and hardened 0-2 tool steel.

TABLE II: Summary of Corrosion Data in 3 wt% Sodium Chloride

COATING	DEP. PARAM.	E p * V SCE	E corr V SCE	PASSIVE RANGE DV V SCE	i corr n amps
Ni	2 A/dm^2	-130	-484	-450 to + 20	4x10E2
316 L	Bulk	>-135	-132	-120 to +380	4x10E3
Brass	Bulk	none	-248	none	7x10E4
Ni-Cr	DC	- 30	-378	-350 to -160	8x10E3
Ni-Cr	50% duty, 12 A/dm2	-78$\underline{<}$	-387	-350 to -100	6x10E3
Ni-Cr	20% duty, 25 A/dm2	-110	-460	-410 to -180	9x10E3
Ni-Cr	20% duty, 20 A/dm2	-110	-392	-350 to - 90	3x10E3

* protection potential

$\leq$ average of 5 samples

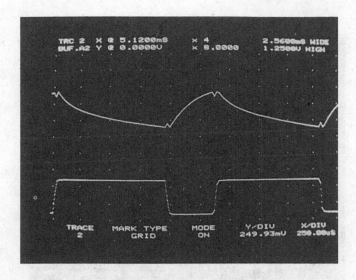

Fig. 1. A typical set of voltage and current
waveforms for the pulsed deposition of nickel-
chromium alloys.

Fig. 2. An SEM micrograph of an etched Ni-Cr alloy
showing layers made up of chromium concentration
gradients.

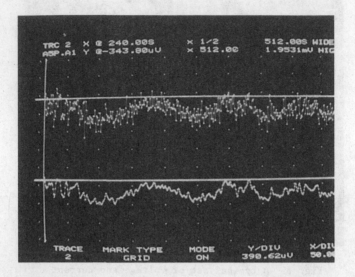

Fig. 3. Voltage oscillation occurring during direct
current deposition (galvanostatic) of a nickel-chromium
alloy. These oscillations are thought to result from
fluctuation in chromium content within the depletion
region. The upper curve is the raw data, the lower
curve has been 5 point averaged.

Fig. 4a. DC, $\sim$ 1 A/dm^2.

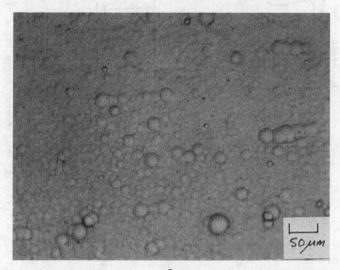

50 μm

Fig 4b. Pulsed, 1.25 A/dm^2, 50% duty cycle, T_{on} = 2.5 msec.
A comparison between direct current deposits and pulsed
current deposits of nickel-chromium.

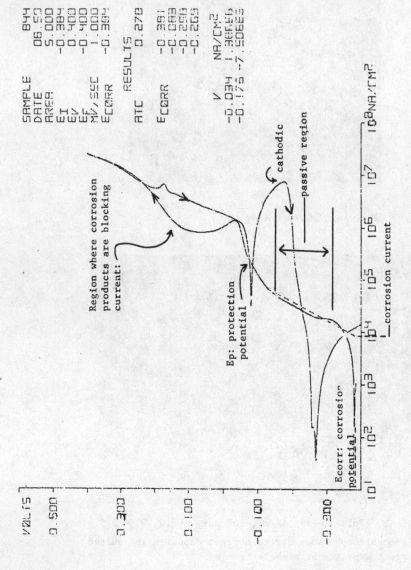

Fig. 5. A "typical" voltammetry curve for a nickel – chromium alloy, showing the protection potential, corrosion potential, and corrosion current.

SESSION II

MEASUREMENT OF COATING PROPERTIES

CHAIRMAN: A. W. RUFF
           NATIONAL BUREAU OF STANDARDS

METHODS FOR (1) TESTING ADHESION, (2) MEASURING THICKNESS, AND (3) MEASURING CRACK AND PORE CONTENT OF COATINGS

H. Hintermann
Laboratorie Suisse de Recherches Horlogers
Neuhatie, Switzerland

Abstract

During the last decade, a new generation of tough, hard, graded composite materials has been developed, i.e., CVD and PVD hard coatings, such as TiC, TiN, Ti(C,N), Al_2O_3, Cr_7O_3, TiO_2, SiO_2, etc., on cemented carbide, steel and Ni-Co alloy substrates. These composite materials are widely exploited industrially in hard coated cutting and forming tools as well as in machine elements such as bearings, gears and general wear parts.

Emphasis will be given to (a) the characterization of the adhesion and mechanical strength of the coating to the substrate material by means of the scratch test, (b) the rapid thickness determination by wear to form a spherical depression, (c) the visualization of cracks and open pores in the coating by electrographic printing (EGP).

Measurement of Residual Stresses in Thin Film Coatings

C.C. Goldsmith and G.A. Walker
IBM General Technology Division
East Fishkill Facility
Hopewell Junction, NY 12533

ABSTRACT

From the viewpoint of structural integrity and long term
reliability, a knowledge of stresses (applied or residual)
is important in the industrial environment. Residual stresses
can arise from a number of different sources such as inter-
metallic formation (volume change) or coefficient of thermal
expansion mismatch (CTE). A variety of non-destructive
techniques for measuring residual stress has been tried
e.g., acoustical, magnetic and X-ray, but to-date only the
X-ray technique has proved practical.

There are a number of ways of measuring strains with X-rays;
however we will examine only those techniques employed in
our laboratory. The most widely used technique is the
polycrystalline method first proposed by Lester and Aborn.[1]
The basic principle behind this technique is simple. One allows
the interplanar spacing of the material to serve as a strain gauge.
This technique was used in conjunction with high temperature
X-ray diffraction to study the stresses in electroless
nickel films plated from a hypophosphite bath.[2] We were
able to show a correlation between stress induced cracking
and Ni_3P precipitation.

We also employ a seldom-used double crystal lattice curvature
technique to measure stresses in thin coatings on single
crystal substrates.[3] Figure 1 shows a schematic of the double
crystal technique with a stressed coating on the single crystal
in position 2. In Figure 1a both crystals are aligned to
obtain Bragg diffraction from a set of crystal planes. In
Figure 1b the second crystal has been translated to a new
position. This translation causes a loss in X-ray intensity
due to the lattice curvature and requires an adjustment in
the angle, Θ, to recover the X-ray intensity. The amount of
adjustment in the Bragg angle Θ, coupled with the translation
distance, l, is a measure of the radius of curvature $(R) =$
$1/\Delta\Theta$. The lattice curvature of the substrate is a direct
result of the stress induced by the film. A major advantage
of this technique, is that the stress measurements can be
made on amorphous films. We have applied this technique
to the study of (amorphous) polyimide films produced from
benzophenone tetracarboxylic acid diethyl ester (BDTA) and
4, 4' methylene dianiline (MDA).[4] Films of this type require
a cure process to obtain the desired properties. To study the
stresses as a function of the cure, an insulated chamber was
added to the camera. The "in situ" measurements show that the

stresses are due to CTE mismatch and that this polyimide film goes through a continuous glass transition.

The above techniques can be combined to allow the determination of the elastic constant $E/(1-V)$.[5] This is done by measuring the strain directly using the polycrystalline technique and substituting the stress measured by the double crystal into the polycrystalline stress equation. The technique was applied to determine the elastic constant for both WSi_2 and $TaSi_2$.

Both the polycrystalline technique and the double crystal lattice curvature technique employ large area X-ray beams. Of course there are many sample configurations that require X-ray stress analysis be determined in small specific areas. We have been successful in employing a specialized X-ray micro-beam technique to determine stresses in areas less than 30μm in diameter.[6] Using this micro-beam technique we were able to stress-map the area between Mo vias screened and fired on Al_2O_3 ceramic and explain the cause of cracking in the Al_2O_3 between vias.

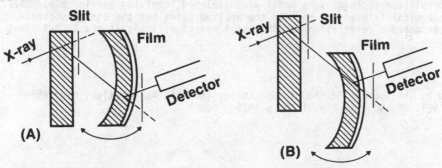

Figure 1. Schematic of Double Crystal Lattice Curavture Technique.

A. First Measurement
B. Second Measurement (second crystal translated distance 1.)

1. H.H. Lester and R.H. Aborn, Army Ordinance, 6, 120-127, 200-207, 283-287, 364-369 (1925).
2. G.A. Walker and C.C. Goldsmith, Thin Solid Films, 53 (1978) 217-222.
3. E.W. Hearn, Adv. X-ray Anal. 20, 273 (1977).
4. C.C. Goldsmith et al, J. Vac. Sci. Tech. April - June (1983)
5. C.C. Goldsmith and G.A. Walker, Adv. X-Ray Anal., 26 (1983).
6. G.A. Walker and C.C. Goldsmith, 16th Reliability Physics Symposium, 56-58 (1978).

PREDICTION OF SALT SPRAY RESULTS FROM PAINT PROPERTIES

F. Louis Floyd
Glidden Coatings and Resins
Strongsville, Ohio 44136

Abstract

A variety of theories have been advanced in the literature to explain the failure mechanism of organic coatings applied to steel substrates. These theories have typically sought to explain performance based on single film properties and have been relatively unsuccessful. This paper presents a progress report on our attempt to predict the corrosion protection behavior of coatings based on the combined relationships among several film properties and salt spray results. The film properties studied were permeability to water and ions, stress-strain behavior, wet and dry adhesion, and electrochemical behavior of the steel substrate in the presence of the liquid paint and the dried film.

The overall salt spray rating had little correlation to any one film property. Using multiple correlation techniques, a model was developed involving barrier properties and electrochemical interaction between the applied paint and the steel substrate. This model produced a correlation of .90, which was significant at the 99.9+% level.

---

* Published by the Journal of the Oil and Colour Chemists Association, Middlesex, England, Vol. 66, No. 11, Nov. 1983, p. 329.

ELECTROFORMED MICROHARDNESS STANDARDS*

D. R. KELLEY, D. S. LASHMORE AND C. E. JOHNSON
ELECTRODEPOSITION GROUP
METALLURGY DIVISION
CENTER FOR MATERIALS SCIENCE
NATIONAL BUREAU OF STANDARDS
WASHINGTON, DC 20234

Abstract: One of the most commonly quoted properties of a material is
its microhardness. Accurate measurements of this property require the
use of precise hardness standards to verify that the testing machine and
testing procedure are accurate; however, it is well recognized that
currently available standards vary considerably in hardness from point to
point. In order to provide more uniform standards, the material used was
prepared by electroforming technology. Two standards are now in production;
one at 125 KHN and the second at 550 KHN. The hardness values are certified
at loads of 0.245, 0.490, and 0.981 Newtons (25, 50, and 100 gram-force)
with both Vickers and Knoop indentors. These electroformed materials have
standard deviation in hardness, particularly at low loads, significantly
better than current commercially available standards. The fabrication of
the new standards and their certification procedures will be discussed.

Key Words: Hardness; knoop hardness; metrology of hardness testing; non-
destructive testing; vickers hardness.

Introduction: Microhardness standards serve as a very important means of
quality control, not only for electrodeposited coatings, but also for many
other metallurgical applications, and can also be used to insure that the
testing machines are operating properly. Presently, the only standards
available are produced by the makers of the testing machines, and these
lack not only a uniform standard of certification of their hardness but
also exhibit a significant variation of hardness across their testing
surfaces. Methods to fabricate microhardness standards that do not have
the above mentioned drawbacks have been developed at the National Bureau
of Standards. By electroforming microhardness standards, very close
control over the three most important variables in electroplating --
current density, temperature, and electrolyte agitation is possible. By
controlling these three variables, it is possible to produce extremely
uniform material which can become a better certified standard. To date,
two standards have been produced, one from a bright copper electrolyte
and another from a bright nickel electrolyte. At present, must available

*This work was sponsored through the NBS Office of Standard Reference
Materials.

standards are produced from cast alloys. The main drawback with this process is that it is extremely difficult to control cooling of the melt. This inability to cool evenly, produces variable grain structure and composition, thus producing a variable hardness across the testing surface. Copper was chosen for the first standard because the nominal hardness of 125 KHN is not only widely used but most closely represents widely tested noble metals. Nickel was chosen because the nominal hardness of 550 KHN closely represents many ferrous metals.

Electroplating, using proprietary brightners, produces grain refinement and grain distribution superior to that achieved with no brightners. This results in a significantly more uniform hardness value across the surface, as well as a bright, smooth surface. Proprietary brightners contain a leveling agent, usually an organic, which is absorbed on the surface of the high current density areas or peaks on the plating surface. This organic inhibits plating on these high current density areas, but allows plating to continue in the valleys, thereby providing a leveling effect. Another characteristic of the bright electrolytes is very good micro-throwing power, e.g., ability to deposit metal in grooves and cracks where these surface imperfections are of a microscopic nature.

Fabrication Procedure: The plating electrolytes used in the electroforming of microhardness standards were commercially available copper and nickel solutions. The electrolyte volume was 100 liters. The anodes were 0.004% phosphorized copper and low sulfur, low cobalt nickel bars placed in anode bags. The power supply was a 15A, 100V constant current source. The electrolyte agitation was provided by filtered air. The substrate used in this process was a 22.5 cm X 45 cm sheet of 1010 polished steel, mounted in a teflon box. Mounting the substrate in a box with the open side facing the anodes provides for a much more uniform current distribution. A uniform current density is required for uniform grain size, essential for this project. Electroforming is stopped when 1 mm of copper or nickel is deposited. Coatings of 500μm of copper on a copper substrate and on a steel substrate showed no significant difference in hardness, therefore the less expensive steel substrate was chosen. The polishing procedure removes approximately 125μm of material from the original coating of 1 mm, thus leaving a substantial coating thickness to prevent any anvil effect from the steel substrate. After the copper or nickel has been deposited, a 2.5 cm strip is removed from all four sides of the plate to ensure thickness uniformity across the plate.

The electroformed plate is cut into 1.35 cm square coupons. All cutting procedures incorporate a high speed, water cooled, carborundum cut-off wheel to minimize deformation. The coupon is then placed in a stainless steel ring 2.5 cm OD, 1.0 cm in height. This ring is then filled with an epoxy medium, used as a mold to enhance uniform polishing. The mold is polished on an automatic system with 600 grit paper, 6μm and 3μm diamond pastes, and a silica suspension to completion. The copper coupons are then flashed with 0.1μm of gold to inhibit tarnish and corrosion.

Certification Procedure: The certification of these test blocks was made in accordance with ASTM B578 and E384. The test machine load-time response was measured using a compression load cell, calibrated with certified weights, to ensure conformity to ASTM E384, Part B. The load cell was used to determine the actual load being applied to the test block during the time of indentation, as well as, determining the dwell time for full load application. The optical measuring system of the test machine was calibrated by a certified stage micrometer. The hardness indentations are made at five different areas of the test blocks and are measured using a 100x objective lens having a numerical aperature of 0.95. The hardness values are certified at loads of 0.245, 0.490, and 0.981 Newtons (25, 50, and 100 gram-force) with both Vickers and Knoop indentors.

Results: Microhardness uniformity was tested using 5 and 25 g-f loads. The 5 g-f load was chosen due to the fact that it is the most sensitive probe of surface properties and was felt to be the lowest load that could be used accurately. Loads less than 5 g-f are extremely sensitive to many variables, e.g., vibration, noise, and air currents. The results obtained from these tests were compared to two commercial standards made of brass and steel. These comparisons are listed in Table 1. Note the standard deviations for these electroformed test blocks are lower than the cast alloy test blocks, indicating improved grain size uniformity and distribution.

Discussion: Electroforming as a materials processing technique provides a substantially more uniform material with respect to hardness than is provided by the available cast alloys used in commercial standards. Hardness standards of 125 KHN and 550 KHN are being produced and are available through the NBS Office of Standard Reference Materials. The loads used are traceable to NBS fundamental mass standards, as are the stage micrometers used to calibrate the optical instrumentation used to measure the hardness indentation.

Future research in the area of microhardness will include the development of three additional standards, lead or tin-lead, with a hardness of 100 KHN or less, electroless nickel with a hardness of 900 KHN, and heat-treated cobalt-phosphorus with a hardness of 1200 KHN.

TABLE 1. A COMPARISON BETWEEN NBS STANDARDS AND COMMERCIALLY AVAILABLE STANDARDS

STANDARD	KNOOP HARDNESS (5 g-f)	
	MEAN	STANDARD DEVIATION
COPPER ON STEEL AS PLATED	131	12.2
COPPER ON STEEL, POLISHED	138	4.8
COPPER ON COPPER AS PLATED	117.2	6.18
COPPER ON COPPER, POLISHED	125.3	4.45
COPPER ON COPPER, POLISHED AND GOLD FLASHED	128.4	3.83
COMMERCIAL STANDARDS		
BRASS	119.4	6.1
STEEL (a)	1103	104
STEEL (b)	1006	87.4
NICKEL ON STEEL AS PLATED	631.3	20.4
NICKEL ON STEEL AS POLISHED	588	17.3

NOTE: THE AVERAGE THICKNESS OF THE COATING EXCEEDED 500 MICRONS

APPLICATION OF THE CHARGE DECAY NDE TECHNIQUE TO SURFACE COATINGS

B. T. Allison, M. K. Tse, J. F. Ramos, N. P. Suh
Massachusetts Institute of Technology
Cambridge, MA 02139

Abstract

A nondestructive evaluation technique has been recently developed for the detection of voids, inclusions, and other subsurface defects in polymeric and other insulating materials. This non-contact technique can also be used to measure the thickness of coatings on metallic substrates. Coating debonds are also detectable.

To apply this method, electrical charge is deposited onto the specimen to be tested and becomes injected into the material. This results in a surface voltage which is measured with an electrostatic voltmeter. Coating thickness variations or debonded regions lead to variations in this potential distribution.

Electrostatic charge decay nondestructive evaluation has been successfully applied in measuring epoxy coating thickness and holds much promise in many other areas.

SESSION III

MARINE COATINGS

CHAIRMAN: H. PREISER
           DAVID TAYLOR NAVAL SHIP RESEARCH
           AND DEVELOPMENT CENTER

ORGANIC COATINGS EVALUATION AND PERFORMANCE PREDICTION
AN OVERVIEW

H. S. Preiser and E. H. Halpern
Coatings Application Branch
Nonmetallics Division
David Taylor Naval Ship Research & Development Center
Annapolis, Maryland 21401

Abstract: Accelerated laboratory test methods are essential for rapid evaluation of protective coatings. Approaches and techniques are available which can cause, detect and follow degradation of paint films under a variety of environmental stresses from its inception to gross failure. These techniques need refinement to achieve reproducibility and reliability of results. Correlation of laboratory tests with field service is essential for ultimate confidence in predicting service life. In this vein, bilge keel panel testing is discussed. The concepts of failure modes and end point of failure are explored. Some techniques not only detect the onset of failure but also provide clues to mechanisms of failure. Several laboratory test methods for accelerated testing of marine paints are described which include: blister testing, rotating drum roughness and wear testing, cavitation damage and adhesion failure, thermal detection of blister inception, and electrical impedence methods for detecting structural and interface changes in coatings systems under environmental stresses. The paper attempts to focus on the theme that accelerated tests can predict performance in service but to have a high degree of confidence they must be correlated empirically with field observations and also be relevant in their mode of failure. If mechanisms of paint degradation could be better understood and their inception rates measured, the time to failure for organic coatings could be predicted with greater certainty.

Introduction: Organic coatings are generally applied over various surfaces primarily to protect the substrate from deterioration by destructive agents in the environment. These coatings can also impart special properties such as aesthetic decorative finishes, provide visual and in some cases thermal information or distort that information for camouflage purposes, provide lubricity for use on sliding surfaces, and in the case of ship hull coatings provide corrosion, erosion and fouling protection in addition to reduced frictional drag. The overall demands made on modern protective coatings are indeed formidable when you consider that coatings systems range from 5 to 100 mils in thickness and are expected to perform in service for extended periods, over five years with minimal repair and touch-up. The engineer and paint technologist is called upon frequently to formulate, modify or select a candidate coating system that would perform its function for long periods of time. Obviously, to qualify a coating based on its real field performance is probably the safest and the most direct course of action, but as a practical matter, these judgements must be made largely from limited, accelerated short-term laboratory data. The underlying theme of this symposium is to review the state-of-the-art in coatings testing technology, focus on some specialized applications, and discuss measurement of properties and failure modes under test conditions. If one could learn how to predict field performance of coatings from their behavior in the laboratory, a great step forward would have been taken towards designing a coating for a given service with a high degree of confidence that it will perform as designed.

Some of the topics covered in Session II, under Measurements of Coatings Properties, are applicable to the development of predictive capability for coatings systems. In this overview, we would like to touch on some accelerated methods being explored at DTNSRDC to evaluate marine coatings performance.

Coatings used on exterior ship hulls are exposed to total and to intermittent immersion in seawater; topside coatings are exposed to sunlight, rain, stack gases, and marine sea spray. Marine coatings are used in conjunction with cathodic protection for underwater hull use. Topside coatings fail by blistering, checking and cracking, loss of adhesion, and by physical and chemical breakdown of the films themselves manifested by loss of color and gloss. Underwater coatings also fail by blistering and loss of adhesion, but since a moving ship is also subjected to high velocity water flow, additional mechanical agents such as wear, cavitation, fatigue and solid and liquid impact cause premature failure. Cathodic protection further stresses coatings by accelerating the transport of water across the paint film by the imposition of a potential gradient (electrosmosis) and increases the pH at the paint surface interface. Antifouling coatings are used over anticorrosion paint systems to discourage attachment and proliferation of biofouling organisms. These topcoats deteriorate by chemical changes caused by the purposeful leaching out of toxic substances. Some antifouling systems must hydrolyze and soften upon contact with the seawater to perform their antifouling function and the resulting spent surface layers are ablated away when the ship is underway. In some cases, the resulting surface roughness is decreased.

Purpose of Test: Laboratory tests are designed to select the best candidate coatings for given service use. Unfortunately, it is not feasible to replicate the field conditions to which the coating will be exposed, nor is it desirable to do so, since the time scale for failure would remain the same. What is necessary is to distort the conditions of exposure to a more severe degree for a comparatively short accelerated time scale to achieve the primary failure mode(s) and correlate such rapid failures with long term field exposure data to extablish predictability by empirical methods. Another approach is to attempt to understand the mechanism of failure and measure initial minute stages or precursors of change in the coating structure and substrate interface by detection with sensitive instrumental sensors. By following the coating/ substrate changes until visible failure occurs, it should be feasible to predict service failure by early detection of microscale changes. Reproducibility of measurements is essential for any analyses to be meaningful. An excellent paper on this general subject was published by Campbell, Martin, and McKnight[1] of the National Bureau of Standards. They recommend that: parametric studies be undertaken of coating response to stresses which are varied singly at high humidity over a range of discrete temperature levels; analytical tools be used for quantitative characterization of materials and material degradation processes; and the mathematical methods (reliability theory) be applied to relate material response to a wide variety of environmental and operating conditions.

Some of the methods being explored at the Center are as follows:

o Accelerated blister testing
o Rotating drum tests
o Bilge keel testing of panels
o Cavitation damage and adhesion failure
o Thermal methods of detecting blister inception
o Electrical methods for evaluating coatings/substrate degradation

Failure Mode and End Point of Failure: Before we discuss briefly these methods for evaluating coating performance and ultimately derive a means of predicting service life, we must understand primary failure modes and define end points of such failures. For underwater service, the most common failure mode of ship bottom anticorrosive coatings is blistering which leads to disruption of film integrity, loss of adhesion and corrosion creep at intact edges progressively lifting the film from the substrate. Surface contamination, ever present on carefully prepared abrasive blasted steel surfaces, are usually the nucleus sites for the start of blister failures[2]. Therefore, when measuring and comparing the blister tendencies of one coating system to another under a series of environmental stresses such as hot distilled water, it is important to establish failure criteria. In previous accelerated hot water blistering tests, the criterion of failure was go or no-go after a period of immersion exposure. In many previous instances, good coatings had blistered due to microcontamination of the surface and therefore had been rejected. We have adopted a blister growth scale which measures the number, size and distribution of blisters as a function of time. Conditions for the test are such that even the so-called best performers from field experience blister within the time frame of the test, hopefully adjusted to occur in the middle of the time scale. We are attempting to establish an end point when the combination of blister size and number results in a loss of adhesion of a given percentage of the panel surface area (presently, arbitrarily chosen between 1 and 5%). At that point, the test is terminated. By arranging the coatings under test on an accelerated time scale, the time to reach the end point is a measure of its blister resistance. The longer the time, the better the coating. These data would have to be correlated with shipboard data on blistering, if predictions of performance in-service is to be attempted with confidence.

For ablative antifouling (AF) coating systems, the primary mode of failure is coating wear as a function of velocity while still retaining its antifouling characteristics. Smoothness is also measured when the coatings are screened for their drag reducing properties. Here again, an end point in the accelerated test for a given series of dynamic/static cyclical exposures at a natural seawater fouling site, is a maximum tolerable wear rate or roughness measure. Coatings that reach these maximums are terminated from the test. The number of exposure cycles to reach this defined maximum wear rate and/or defined roughness height is a measure of the suitability of these coatings for shipboard service.

Another failure mode is loss of gloss and color retention under intermittent weathering and UV exposure at the maximum wavelength of 310 nm. By setting limits on color changes and gloss changes that are acceptable, these test parameters can be measured to the failure end points. The time to cause such failure is the index of coating superiority. These data also would have to be correlated with similar coatings exposed to field conditions to establish capability in service on the basis of laboratory accelerated testing.

Methods of Testing under Investigation

Blister Growth Test: This is a modified hot water test specified in MIL-P-24441[3].
It was found during preliminary work that the Navy Epoxy Paint, Type I (solvents do not conform to California Rule 66 for air pollution) can be blistered within two weeks if distilled water is used in lieu of tap water, the temperature is increased from 180 to 195°F and the coating thickness is at least 6 mils dry film thickness. Figure 1 shows the proposed diagramatic circuit of the test apparatus. The water is circulated through a still and make-up feed is provided by an ion exchange unit and demineralizer to maintain a high quality distilled water in contact with the paint

surface. Figure 2 shows the difference in the blister growth to end point of Type I, Navy Epoxy (non-complying) paint with Type II in which the solvents have been modified to comply with current air polution laws. By keeping the surface preparation techniques constant during panel preparation and clearly marking the major inclusions or contaminants, it appears to be feasible to separate coating performance on the basis of blister growth to an arbitrary but carefully chosen end point. These data will be generated for a large variety of commercially available equivalents of MIL-P-24441 to separate the poor performers from the good. These data will be correlated with bilge keel panels mounted on operating ships which are described later. It should be mentioned that these blister tests will also be run in hot salt water under cathodic voltage gradients of -0.75 to -1.5 V, referred to the Ag/AgCl half-cell and corrected for temperature to determine if blister growth can be further accelerated under these conditions.

Rotating Drum Test: The rotating drum apparatus was designed and constructed by Miami Marine Research, Inc., to test antifouling coatings under dynamic conditions, while being fully submerged on site (Biscayne Bay, Miami Beach, Florida). The drum is fabricated from PVC and each test panel is attached to the outer surface with four nylon countersunk bolts, thus avoiding electrolysis problems which would occur with metallic fasteners. Figure 3 shows the drum apparatus partially submerged, being filled with water in its hollow interior to reduce its buoyancy, after the test panels were installed on its outer circumference. The test panels are subjected to 45 days of continuous rotation at peripheral speeds up to twenty two knots. Figure 4 shows a theoretical relationship between drum rotational speeds and equivalent ship speeds to achieve similar hydrodynamic shear values. At the end of the dynamic period, all test panels are inspected and photographed, then placed on static exposure mode for an additional 45 days to observe fouling accretion.

The failure end points in these tests were threefold:
   a. If the panels fouled in the dynamic cycle or failed to ablate any fouling accumulated during the static exposure portion of the cycle.
   b. If the wear rate during rotation exceeded 2 mils per month.
   c. For certain applications, if the roughness increased beyond 600 microns mean apparent amplitude (MAA as measured by a BSRA Hull Roughness Analyzer[4]) or measured with a stylus instrument at 0.3" cut off.

These tests were used as the basis for coating selection for full scale shipboard tests in order to establish a qualified products list of approved long term AF ship bottom coatings. These data will also be correlated with bilge keel panels, described below.

Bilge Keel Tests: Panels are mounted on welded studs arranged on the bilge keel of a fast frigate. The panels, coated with candidate coatings, are butted with minimal clearance edge to edge, and are thoroughly grounded to the hull in order to be exposed to the effects of the shipboard cathodic protection system. The panels have special zinc anode grommets to insure good electrical contact to the hull through the stud mounting. Figure 5 is a schematic detail of the panel arrangement. Sufficient numbers of panels of each candidate coating system are available to obtain data with a high degree of confidence. The panels are removed by divers on a six-month schedule and examined for blister growth, wear, AF resistance and roughness. The results are correlated with the accelerated blister growth test and rotating drum test, described previously.

Cavitation Damage and Adhesion Failure: Cavitation damage to coatings is a subject of high interest at the Center. A vibratory horn apparatus, similar to ASTM G-32-72, was modified to permit cavitation intensities experienced in the field to be applied to painted substrates as shown in the diagramatic sketch of Figure 6. When coating destruction took place in very short time intervals, these systems were rejected. Using high energy absorbing elastomers such as specially formulated urethanes, these coatings were able to sustain visible damage for considerable extended periods (20 hours). Energy absorbed in such coatings generally weakens adhesion bonds which could lead to its ultimate peeling at resulting discontinuities on hydrofoil surfaces. A series of hesiometer measurements[5] were made as a function of water immersion time with and without cavitation exposure. The results are shown in Figures 7 and 8. These data are now undergoing correlation of coating performance in a high speed water tunnel which will model cavitation flow conditions. Another technique being considered is acoustic emmision, which will be used to try to measure discrete acoustic events occuring in the coating structure and at the interface early in the cavitation damage cycle. It is postulated that these changes are precursors to detecting early coating failure and thereby offer a potential means for rapid screening of coatings. This technique will be discussed in detail in a paper by S. Basu and G. S. Bohlander in this session of the symposium.

Thermal Methods of Determining Blister Inception: Differential thermal conductivity methods (thermography) can determine early onset and location of blisters on painted substrates. Long term (10 years) paint requirements can be met by anticorrosive paint systems that are blister resistant. Methods for detecting early incubation of blisters are needed to develop these anticorrosive paints. Thermographic methods are now being used to study blisters at NBS[6] and adhesion of flame sprayed metal coatings on substrates at DTNSRDC. Differences in thermal conductivity between water, air and paints are sufficient to pick up hot spots on heated painted steel panels. A 10° C differential between the panel and the ambient temperature has been sufficient to pick up incipient blisters. Increasing the temperature differential will increase the sensitivity of the system but would be limited by the thermal destruction of the paint. Additional increases in resolution of the infrared sensors are needed to detect pre-blister areas under the initial applied paint films. The basic approach in using thermography to detect blisters will be to stress the samples in an accelerated environment that will produce blisters as a function of time until a failure end point is achieved. The correlation of incubation time of blister with gross failure may lead to rapid identification of blister resistant paints. This rapid feedback of blister information could be used to improve selection and formulation of paints and related surface preparation parameters to achieve a ten year service life.

Electrical Methods for Evaluating and predicting Paint Performance: Electrical methods are part of the program of non-destructive test methods for detecting early deterioration of paint films. These methods can detect changes in internal paint properties and in the region of the film/substrate early in the performance cycle to serve as indicators and predictors of long term performance. These methods are potential diagnostic tools for coating development and qualification[7]. ONR is currently funding some theoretical work which deals with impedance measurements of paint films on metal substrates[8]. These measurements contain chemical and structural information of the coatings through the capacitance and inductance components of the impedance. The capacitance term contains chemical information through changes in the dielectric constant. For example, when water penetrates into an organic film, even without swelling, the dielectric constant, or dipole configuration changes resulting in a shift in capacitance as shown in Figure 9. The same water could also cause morphological changes due to swelling or other internal stresses, as well as instability at the film/substrate interface which would affect the geometric term of the capacitance equation, thereby resulting in a change in the total capacitance. The separation of these effects in the capacitance measurements would lead to insights on mechanism of degradation. The same holds true of inductance measurements which also contain chemical information due to the magnetic permeability term and structural information due to equivalent coil geometry term. Intermediary changes in chemical reactions are normally paramagnetic leading to a new organic species in the diamagnetic state which is shown schematically in Figure 10. Structural information related to spacing and intertwinning of polymer chains may be detected in the geometric term of the inductance measurement. It can be seen from a conceptual viewpoint that the separation of capacitance and inductance effects, both chemically and geometrically could lead to a more complete understanding of the mechanism of coating degradation with time under environmental stress[9]. This technique is scheduled for further study at the Center.

Conclusion: This overview was meant to just scratch the surface on the possibilities of developing accelerated laboratory tests which would predict coatings performance in service. When tied to mechanistic studies, the possibilities of designing a paint system for a specific service and performance life appear to be within reach.

The views expressed in this paper are solely those of the authors and do not necessarily reflect the official position of the Naval establishment at large.

References:

1. Campbell, P.A., J.W. Martin and M.E. McKnight, "Short Term Evaluation Procedures for Coatings on Structural Steel," NBS Technical Note 1149, U.S. Dept. of Commerce, Sept. 1981.
2. Mumger, C.G., "The Coatings of Contaminated Surfaces," 3210 Sage Road, Fallbrook, CA 92028
3. MIL-P-24441, Paint, Epoxy-Polyamide, General Specification for, Naval Publications and Forms Ctr., 700 Robbins Ave., Philadelphia, PA 19111, Attn: Dir. 4ND.
4. BSRA Technical Services, "Hull Roughness Analyzer Gauge and Hull Survey Service," Wallsend Research Station, Wallsend Tyne and Wear, NE286UY, Tel. 0632-625242, UK.
5. Asbeck, W.K., "Measurement of Adhesion in Absolute Units By Knife Cutting Methods, The Hesiometer," IX Patipec Congress (1968).
6. McKnight, M.E., and J.W. Martin, "Non-Destructive Corrosion Detection under Organic Films using Infrared Thermography," SAMPE Fall Mtg, 1982, Washington, DC.
7. Leidheiser, H.E., Jr., et al, "Corrosion Control through a Better Understanding of the Metallic Substrate/Organic Coating/Interface," Lehigh University, Bethlehem, PA, Contract #N00014-79-C-0731, 3rd Report, 1 December, 1982.
8. Beck, T.R.,and R.T. Ruggeri, "Determination of the Effect of Composition, Structure and Electrochemical Mass Transport Prop rties on Adhesion and Corrosion Inhibition of Paint Films," 3 reports, Electrochemical Corp., 3935 Leary Way, NW, Seattle, WA 98107, under Contract #N00014-79-C-0021, available as AD A095-293, AD A113-975, NTIS, US Dept. of Commerce, Springfield, VA 22161.
9. Halpern, E.H., and J.C. Sherman, "Resonant Frequency of an Oscillator used to Determine the Thermal Stability of Magnet Wire and Varnish," 1970 Annual Report, Conference on Electrical Insulation and Dielectric Phenomena, National Academy of Sciences, Washington, DC, 1971.

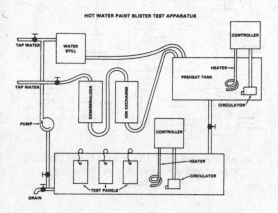

Figure 1 – Hot Water Paint Blister
Apparatus

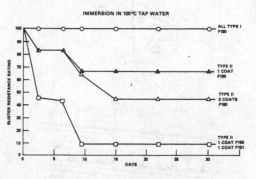

Figure 2 – Blister Rating vs Time for
Type 1 and Type 2 F-150 Navy Epoxy
Coatings

Figure 3 – Rotating Drum Apparatus

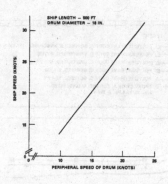

Figure 4 – Drum Speed to Give Equivalent
Average Shear Force on a 500 Foot Ship

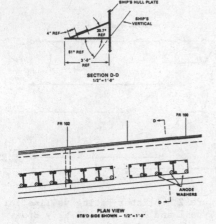

SECTION D-D
1/2" = 1'-0"

PLAN VIEW
STB'D SIDE SHOWN — 1/2" = 1'-0"

Figure 5 – Schematic Diagram of Bilge
Keel Panels for Shipboard Testing

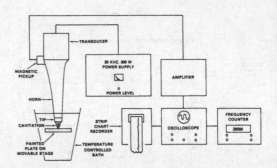

Figure 6 – Block Diagram of Modified
Vibratory Apparatus for Testing
Paints

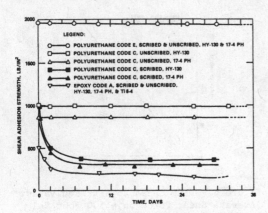

Figure 7 – Immersion Time vs Adhesion
Strength for Several Candidate
Coatings

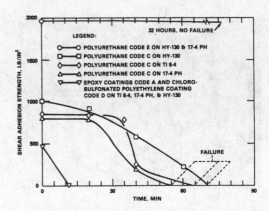

Figure 8 – Shear Adhesion Strength vs
Cavitation Time for Several Candidate
Coatings

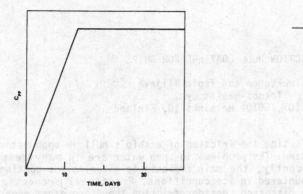

Figure 9 – Film Capacitance vs Time
Showing Water Uptake for a
Typical Film

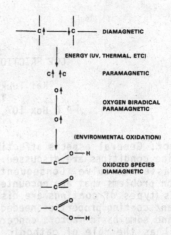

Figure 10 – Typical Oxidation from
Diamagnetic State Through
Paramagnetic State to
New Diamagnetic State
of Organic Films

71

LOW FRICTION HULL COATINGS FOR SHIPS

Kai Tuukkanen and Tapio Viljava
Teknos-Maalit Oy
P O Box 107, 00101 Helsinki 10, Finland

Abstract: General aspects affecting the friction of a ship´s hull in open water and under ice conditions are discussed. The problems in open water are in many respects much easier to solve. Consequently, the main emphasis is given to the coating and friction problems that are encountered in ice conditions. Frictional properties of various types of coatings are discussed. Besides friction there are other even more important coating properties needed for an "in practice" low friction hull coating. These and some basic factors concerning the application of the coating are discussed, as well as the role of cathodic protection. It is clearly proved by practical examples that major economic advantages can be and have been achieved by the use of a special low friction ice resistant coating.

Key words: Inerta 160; friction; hull coating; icebreaking

INTRODUCTION

It is a well-known fact that an increase in hull roughness will considerably increase the fuel consumption. The hull roughness is due to
- mechanical damage
- corrosion
- fouling
- maintenance practice leading to uneven film thickness

Consequently, a "low friction hull coating" should be able to prevent
- mechanical damage (leading to corrosion)
- corrosion
- fouling

It should also be able to smoothen rough, corroded surfaces. You must be able to apply it in temperatures and relative humidities which exist or can be reasonably created in shipyards and drydocks and by using modern, efficient application techniques available. In addition, the frictional characteristics of the coating material should be as good as possible.

HULL ROUGHNESS

Hull roughnesses in practice have been measured and discussed by several authors /6, 11, 12/. The annual roughness increase varies considerably from ship to ship. However, the figures given by Townsin /12/:
        25 microns per annum on well maintained ships
        80 microns per annum on poorly maintained ships

40 microns per annum as an average
correspond quite well with the general experiences.

According to Hacking /6/, an increase of 2-3% in engine power is required to maintain
the original service speed for each 25 micron increase in mean apparent amplitude of
the underwater hull. From these figures the importance of hull smoothness is quite
obvious.

RESISTANCE IN OPEN WATER

In normal open water transit a ship must overcome the hydrodynamic friction, which
depends on the smoothness of the hull and, of course, of the hull shape. The
frictional characteristics of the coating material itself are overruled by its surface roughness.

The mechanical and corrosion resistance properties needed in open water are fairly
easy to achieve with modern anticorrosives. By using on top of them a long-lasting
antifouling based on a self-polishing copolymer (SPC) system it is possible to
achieve, besides good antifouling properties, a reduction in surface roughness during
the ship's time in service, as an additional advantage. Due to this "double effect"
considerable savings in fuel consumption can be achieved.

RESISTANCE IN ICE CONDITIONS

Until the mid 70´s there was no coating system known, which could withstand the
tremendous impact of ice. Consequently, e.g. icebreakers were not coated at all or
the coating was done for temporary corrosion protection, since the paint coat was
worn off during the first hours of ice navigation. From this it is obvious that quite
exceptional resistance properties are needed before a coating performs in practice as
a "low friction hull coating".

In transit through ice a ship has to break the ice, displace the ice blocks and overcome the friction between the hull and ice. The area of heaviest impact and friction
is along the water line. The width of this area is a function of the hull form, the
thickness of ice and the ship´s speed.

For obvious reasons fouling is no problem in ice conditions.

FACTORS AFFECTING FRICTION

Several scientists have studied the frictional behaviour between ice and other materials. One of the most comprehensive studies is made by Calabrese /2/, who has been
involved in many thorough investigations in conjunction with friction and ice conditions. The results and conclusions found in the literature tend to be contradic-
tory. This must be considered as a proof of the complexity of the phenomenon in
question.

In general, organic polymers used as binders in coatings have lower friction coefficients than steel. Examples of these polymers are polyolefins such as polythene and
fluoropolymers such as Teflon. Unfortunately, polymers having the lowest friction
coefficients do not possess the basic coating properties needed.

When discussing the friction coefficients, it is doubtful if the values given represent pure material characteristics without the influence of variations in surface roughness and profile, properties of ice and, in the transition region from static to kinetic friction also the speed. In addition, surface pressure, water layer on ice, temperature, humidity etc. also influence the friction. For example snow on the ice increases the friction considerably. At low temperatures the friction of snow is almost as great as that of sand. Consequently, it is difficult to judge the influence of various factors on the friction between a ship's hull and ice. However, based on the experiences obtained so far it is believed that compared to other surface properties, the importance of hull roughness is crucial. Thus, if the hull can by some means be made smooth, the practical benefit of a lower friction coefficient is only marginal. In other words, of two coatings the one which is smoother and remains smoother is the best low friction hull coating.

INERTA 160 AS A LOW FRICTION ICE RESISTANT COATING

As shown above, the frictional properties of an ice-going ship are dependent on whether the surface remains smooth, that is whether the coating endures the strains of ice transit and protects the hull from corrosion. This being the fact, the tests that are decisive when developing a coating are not measurements of friction in laboratories, but full-scale tests with icebreakers in severe ice conditions. The only coating that has passed these endurance tests so far is INERTA 160.

The first full scale trials were made during the years 1972-1974 and they were continued on Finnish, US and Soviet icebreakers. The details of these tests have been published earlier /13, 14/.

From the very beginning the coating proved to have quite exceptional resistance properties and it was also proven in the first test with two Finnish coast guard cutters that considerable savings can be achieved due to reduction in friction.

A summary of some of the earlier experiences with INERTA 160 is given in Appendix 1.

APPLICATION CONDITIONS AND TECHNIQUE FOR INERTA 160

Requirements on painting conditions are the same as for epoxy paints in general. The surface temperature must not be lower than +10°C; the recommended minimum temperature is +15°C. The relative air humidity at the surface temperature must not exceed 80%.

The surface to be coated should be blast cleaned to Sa 2 1/2 or Sa 3. Shop primer coats if present must be thoroughly removed, since their adhesion is markedly poorer than that of INERTA 160. The effect of surface preparation on adhesion is clearly shown by the following pull-off adhesion tests. Epoxy shop primed test plates were blast cleaned to different grades and painted with INERTA 160. The preparation grades and corresponding pull-off strength values are as follows:

Preparation grade	Pull-off strength/MPa
No blasting	11.5
Light sweep blasting	16.2
Grit blasting to Sa 2	23.1
Grit blasting to Sa 2 1/2	25.1
Shot blasting to Sa 3	27.5

INERTA 160 and the more thixotropic INERTA 160 FILL are applied by a two-component spray. On pitted steel, INERTA 160 FILL should be smoothed by putty knife immediately after spraying. A two-component spray gun with static mixing system is used for application. In Finland there are over 40 spray guns of this kind in use; ship yards and bigger painting contractors have their own apparatus. Differently from the normal types of spray guns, it is vital that two-component guns are always in good condition so that proportioning and mixing errors can be avoided. Regular checking of the functions of the spray gun is an easy and speedy job, and it is adopted as a routine by the Finnish ship yards.

INERTA 160 system alternatives for new and pitted steel are given in Appendix 2.

INERTA 160 TODAY

Today INERTA 160 is with excellent results used on all icebreakers in the Finnish coastal service. Also several other types of vessels, e.g. ice-going tankers, have been coated with INERTA 160. Older ships with pitted hulls are coated with a system of INERTA 160 FILL and INERTA 160. Also ships operating mostly in open waters, e.g. FINNJET, the fastest car/passenger ferry in the world, are coated with INERTA 160.

In the Soviet Union nine Finnish-built icebreakers coated with INERTA 160 were inspected after 1 to 4 years of service /4/. After one or two years operation in ice, 85 to 90% of the coating was still intact. On IB YERMAK, 85% of INERTA 160 was still without damage after four years. On icebreakers operating in rivers, where they have to break sand-bearing ice in shallow water, the damage was somewhat greater.

Because of previous good results the U.S. Coast Guard have had the hulls of several icebreakers coated with INERTA 160, including their new series of 43 metre domestic icebreakers.

The latest experiences come from IB NORTHWIND /3/, that was coated after the 1980 season. In this phase the aim was to protect the hull from corrosion, since the renewal of rapidly corroding plates and welds was costly. The anti-corrosion effect as desired was achieved. In 1981 from October to December NORTHWIND operated in severe ice conditions in the Greenland Sea. An inspection after the expedition showed that more than 95% of the coating had remained intact.

INERTA 160 AND CATHODIC PROTECTION

INERTA 160 has also made it possible to use new propeller material in icebreakers. Up till now the propellers have been made of cast iron containing 4% nickel, which is very susceptible to electrochemical corrosion. Due to corrosion, the dimensions of this kind of propellers change rapidly resulting in lower operating efficiency.

A propeller that maintains its dimensions and efficiency can be made of stainless martensite steel that contains 13% chrome and 4% nickel; this material is prone to pitting, though. This problem as well as corrosion of the hull can be solved by cathodic protection. The following conditions, which demand much of the coating, were set as a basis for investment calculations by the Finnish Board of Navigation /7/ when they were planning to introduce cathodic protection and a new propeller type:

1. No more than 50% of the hull coating should be worn off within five years.
2. It must be possible to install the anodes so that they remain fixed throughout the five years of operation.

75

3.   The interval between drydockings must be two years, and it should be possible to postpone the dockings till the off season.

Many years experience has shown that these requirements are clearly exceeded when INERTA 160 is used for coating and filling. Inspections have revealed that the annual damage on INERTA 160 amounts to no more than 1 to 5%.

A calculation based on the above-mentioned supposition of hull coating costs of a Tarmo class icebreaker /14/ shows that as an investment the INERTA 160 system is more advantageous than conventional systems, and if fuel savings are taken into account, the financial gain is substantial.

INFLUENCE OF INERTA 160 ON FUEL CONSUMPTION

The fuel savings achieved by the application of INERTA 160 can be calculated directly from the reduced need of energy. Savings have been achieved both in ice service and when trading in open waters.

Oy Wärtsilä Ab, Helsinki Ship Yard has made a calculation of the savings obtained in the annual propulsion energy consumption on MS TURELLA by the application of INERTA 160 /5/. TURELLA is trading between Turku and Stockholm, and the ice conditions involved are relatively easy. Savings achieved during the half-year open water season are estimated to be about 6% and during the winter season about 8%.

Wärtsilä has also made comparative tests on ice resistance with two sister vessels, the Soviet icebreakers KAPITAN DRANITSYN and KAPITAN NIKOLAEV /8/. The hull of DRANITSYN was coated with INERTA 160, whereas NIKOLAEV was left uncoated. The reduction in ice resistance measured resulting from the use of INERTA 160 was 18%.

The U.S. Coast Guard observed that the use of INERTA 160 on their IB NORTHWIND improved the ship's efficiency and reduced the fuel consumption considerably. Before, the effect of two engines at both shafts had been required for icebreaking. After the application of INERTA 160, two engines instead of four were enough in icebreaking service. Compared with the previous fuel consumption in icebreaking service, a 4,000 gallons saving per day was recorded. In addition, it was observed that about 1,000 gallons less per day was consumed when operating in open water. All together, this meant a 15% saving in fuel per nautical mile. Besides fuel savings, benefit was drawn from the possibility of employing the engines more rationally. On one hand, the need for maintenance of the engines decreased as only one engine instead of two per shaft was employed, and on the other hand more time was available for maintenance of the non-operating engine, while the other was running. All in all, the coating was estimated to pay for itself in mere fuel savings in two years.

To make savings in continuous use possible, the coating must, of course, stay smooth during navigation. The following hull roughnesses measured from Finnish ice-going coastal tankers show that the increase in average hull roughness (AHR) on ships coated with INERTA 160 is minimal:

Ship	AHR	Age of coating measured
MT SOTKA	160 microns	7 years
MT LUNNI	237 microns	7 years
MT UIKKU	160 microns	6 years

The higher figure of MT LUNNI is probably due to the fact that the ship was fouled and only partially cleaned before the roughness measurements were performed.

The case of MS ESSO FINLANDIA shows the rapid increase in roughness with conventional coatings and on the other hand the smoothening effect of INERTA 160. ESSO FINLANDIA was built in 1981 and coated with a chlorinated rubber system. The AHR was measured in March 1983 to 205 microns. Afterwards the ship was coated with INERTA 160 and this decreased the AHR to 125 microns.

All above-mentioned measurements were made with a BSRA hull roughness analyser.

In addition to lowering the AHR, INERTA 160 reduces friction also by making the profile less sharp, which is not recorded in AHR values.

CONCLUDING REMARKS

The new era in the coating of ice-going vessels has now lasted 10 years. During this period about 300 ships have been coated with INERTA 160. The results obtained have been favourable both technically and economically:

-        INERTA 160 exhibits good performance even in the severest ice conditions and protects the hull from corrosion. Thereby the need for dockings decreases and expensive repairs of hull plating are avoided. The application of INERTA 160 makes it possible to apply cathodic protection to the ship's propellers and hull, and thus corrosion can be practically fully avoided.

-        The application of INERTA 160 reduces hull roughness and friction, and in this way also ice resistance. This leads to better icebreaking efficiency in extreme conditions, longer radius of activity, and appreciable fuel savings. Savings have also been proved when operating in open water.

Economic advantages achieved through the use of INERTA 160 are indisputable. Either the anti-corrosion protection or the fuel savings alone is sufficient to make the use of the coating profitable. Both benefits put together, INERTA 160 pays for itself in one or two years depending on the proportional share of ice navigation.

Apart from ships, INERTA 160 has been used with success on other structures which require a coating with special properties, such as good adhesion and mechanical strength at variable temperatures. INERTA 160 has been applied e.g. to railway freight cars carrying ore and other bulk goods. In the processing industry and power stations there are many demanding structures that have been successfully coated with INERTA 160, e.g. power station sluices. A chapter by itself are nuclear power plants which set strict requirements on the durability of coatings. In nuclear power plants INERTA 160 has been used for coating steam pipes, deionized water tanks, sea water pipes and equipment of heat transmission systems, etc.

So far INERTA 160 does not have any competitor in its own class. Although INERTA 160 is called "a solventless epoxy coating" according to its binder type, it should be noted that the results of the performance of INERTA 160 cannot be extended to apply to other coatings of similar type. Practical trials evidence that the results achieved with other solventless epoxies cannot be compared with the performance of INERTA 160.

REFERENCES

1. BYRNE, D., Hull Roughness of Ships in Service. Meeting on the North East Coast Institute of Engineers and Shipbuilders 1979. 8 p.

2. CALABRESE, S.J., PETERSON, M.B., LING, F.F., Low Friction Hull Coatings for Icebreakers, Phase I/1. USCG Report NTIS 784 361. New York 1974. 94 p.

3. EGAN, D.M., Recent Advances in Energy Savings Achieved in Icebreaker Operations. SNAME Spring Meeting 1983. 9 p.

4. GOMAN, G.M., KLIMOVA, V.A., SOLODOVNIKOVA, E.G., Corrosion Condition of Ice-Breaker Underwater Hull. Ministry of Shipbuilding Industry (USSR) 1982. 7 p.

5. GORDIN, S., PULLIAINEN, J., The Effect of INERTA 160 on the Energy Consumption of MS TURELLA on the Route Turku-Stockholm (in Finnish). Wärtsilä Arctic Design and Marketing 1981. 3 p.

6. HACKING, H., The Economy of Smooth Hulls. International Ship Painting and Corrosion Conference 1974. 7 p.

7. HARJULA, A., Recent Developments in Underwater Hull Protection in Finnish Ice-breakers. Finnish Board of Navigation. Helsinki 1982. 7 p.

8. JUURMAA, K., WILKMAN, G., The Ice-Breaking Tests of IB KAPITAN DRANITSYN and IB KAPITAN NIKOLAEV on River Yenisey (in Finnish). Wärtsilä Arctic Design and Marketing 1981.

9. MAJOR, R.A., GULICK, R.W., CALABRESE, S.J., Abrasion-Resistant Coatings and Their Application to Ice Transiting Ships. Presented at Chesapeake Section of SNAME 1978.

10. MÄKINEN, E., LAHTI, A., RIMPPI, M., Influence of Friction on Ice Resistance, Search for Low Friction Surfaces. Ice Tech 75. Montreal 1975. 10 p.

11. TOWNSIN, R.L., The Economic Consequences of Fouling and Roughness. Oyez International Business Communications Ltd., Marine Fouling Seminar. London 1979. 3 p.

12. TOWNSIN, R.L., BYRNE, D., MILNE, A., SVENSEN, T., Speed, Power and Roughness: The Economics of Outer Bottom Maintenance. The Royal Institution of Naval Architects 1980. 17 p.

13. TUUKKANEN, K., TALLGREN, H., Recent Experiences of Bottom Protection on Ice-going Vessels. International Ship Painting and Corrosion Conference. Monte Carlo 1978. 4 p.

14. TUUKKANEN, K., VILJAVA, T, Unique Ice Resistant Coating; Practical Experiences, Economic Aspects. POAC 83 Conference. Helsinki 1983. 10 p.

APPENDIX 1

EARLIER INERTA 160 MILESTONES

MS SILMÄ	1974	30% reduction in ice resistance; INERTA 160 ok
IB MURTAJA	1974	8,000 nautical miles and 900 hours of navigation in ice; INERTA 160 ok, other coatings worn off
IB YERMAK	1975	21,500 nautical miles, more than 3,000 hours in icebreaking service; INERTA 160 ok, other coatings totally disappeared
IB SISU IB URHO	1976 1976	INERTA 160 in good condition while competitive coating on sister ship IB URHO was badly damaged and had to be recoated completely in the waterline areas
IB URHO	1978	Due to the above experiences the Finnish Board of Navigation decided to coat IB URHO plus all other icebreakers with INERTA 160
IB GLACIER	1977	Starboard bow and a section stretching over the bow rib coated with INERTA 160; port bow coated with ZEBRON
	1979	INERTA 160 in excellent condition; ZEBRON totally disappeared

APPENDIX 2

INERTA 160 SYSTEM ALTERNATIVES FOR NEW AND PITTED STEEL

1.   **New steel**
        1 x 500 - 600 microns INERTA 160

2.   **Old, evenly pitted (0.5 - 1 mm) surface. Also new steel in case the surface
        will be subjected to heavy abrasion:**
        2 x 500 - 600 microns INERTA 160

3.   **Pitted (1 - 3 mm) steel and around anode areas:**
        1 x 1.0 - 1.5 mm INERTA 160 FILL
        1 x 500 microns INERTA 160

4.   **Heavily pitted (3 - 5 mm) steel:**
        1 x 2 - 5 mm INERTA 160 Putty (= 1 part INERTA 160 to 1.5 parts Fillite by
        volume)
        1 x 500 microns INERTA 160

5.   **Special uses or filling up, e.g. installation of anodes:**
        1 x 5 - 20 mm INERTA 160 Putty (= 1 part INERTA 160 to 2.5 parts Fillite by
        volume)
        1 x 500 microns INERTA 160

The surface temperature should be 15 - 40°C. Intervals between coats of different
systems vary from 2 to 12 h depending on the temperature. INERTA 160 FILL and INERTA
160 Putty can be overcoated immediately after completed application.

The cured coating should be roughened by sweep-blasting before a further coat is
applied. Otherwise the applied coating will detach due to poor inter-coat adhesion.

INERTA 160 FILL

INERTA 160 FILL can be applied with a two-component spray. In order to achieve a
uniform finish with spray nozzles 0.026 - 0.031", INERTA 160 FILL must be heated up
to 50 - 70°C, and the pressure of the pump must be about 300 bars. Low pressure or
cold paint gives an unsatisfactory spray pattern.

On pitted surfaces INERTA 160 FILL should be smoothed with a putty knife. The smoothing must be done immediately in connection with the spraying. The surface smoothed
with knife must always be overcoated with either INERTA 160 FILL or INERTA 160.

INERTA 160 Putty

INERTA 160 Putty is made by mixing 1-2.5 parts of Fillite micropearls, grade 200/7
(Fillite Runcorn Ltd., Runcorn, England), to 1 part of ready mixed INERTA 160.
INERTA 160 Putty is applied by hand with a putty knife.

THERMAL-SPRAYED COATINGS FOR CORROSION PROTECTION

H. Herman and H. Bhat
State University of New York
Stony Brook, New York 11794

Abstract

Thermal-sprayed coatings can provide highly effective protection for a wide range of substrates in aggressive mechano-corrosive environments. Flame, electric-arc, and plasma spray techniques are readily available for forming both electrochemically active (e.g., zinc or aluminum on steel) and barrier (e.g., Ni-based alloys) coatings for protection in an enormous variety of industrial applications. There remain, however, deficiencies in the use of these coatings such as porosity and, frequently, limited adhesion strength. Furthermore, there are no readily effective NDE techniques which are available in the field for quality assessment. These limitations are fortunately being overcome. For example, high temperature corrosion resistant alloys are being plasma sprayed under controlled environments or in low pressure chambers, producing coatings of greatly enhanced density and adhesion strength, with significantly improved mechanical and corrosive behavior. Also, post-spray laser treatment can be employed to seal surface porosity, yielding a dense surface alloy with excellent corrosion and oxidation resistance.

A review will be presented of selected recent advances in thermal spraying and in materials processing methodology. Emerging NDE testing concepts will also be discussed.

The overall thrust will be to demonstrate the practicabilities of thermal spray technology for solutions of actual industrial problems.

COATINGS FOR CORROSION CONTROL OF NAVY SHIPS

Vincent J. Lanza
Metco Inc.
1101 Prospect Avenue
Westbury, New York 11590

Abstract: For over ten years, a great deal of effort has been spent
by both the Navy and industry to reduce corrosion aboard ships. This
effort was in the utilization of the metallizing process applying an
aluminum coating.

This coating was applied by feeding a pure aluminum wire into an oxygen
and acetylene flame and atomizing air. The air takes the molten particles and propels them to the substrate. This process was clearly
identified and published in the AWS 19 Year Corrosion Report.

This process was demonstrated to the Navy in 1971 and resulted in tests
on the launch and recovery areas aboard carriers. The results of these
tests were very favorable and reported in NAEC-ENG-7839 report, dated
17 Dec. 1973.

This paper will endeavor to identify the approved application, tests
and specifications generated after the initial tests in 1971. We would
also attempt to describe cost savings in manpower as well as equipment,
utilizing this corrosion control process.

One of the major problems facing our Navy is the protection of our
fleet from corrosion. We have demonstrated that most of these problems
of corrosion can be solved by applying a thin flame sprayed coating of
aluminum to the substrate. This coating was selected by the tests performed over a 20 year period by the American Welding Society (AWS).

In 1971, a wire sprayed coating of aluminum was applied to a Launch and
Recovery Steam Valve aboard a Navy carrier. The lagging was removed
and a thin coating of wire sprayed aluminum was applied (approximately
6 to 8 mil thick). The lagging was then replaced and then, after one
year of service, it was removed and the valve checked. The valve was
found to be in excellent condition with no degradation and expected to
have a service life of a minimum of five years.

This was the initial step in directing the Navy to initiate a policy of
using a wire sprayed aluminum coating for external preservation of
steam valves vis-a-vis heat resistant aluminum paint. See Report
NAV SEA S6435-AE-MMA-010 NAV AIR 50-20-1. This report supplies a
general description of the aluminum coating system, equipment and
procedures. Coatings were applied to the USS Schofield (FFG 3) for
evaluations in 1975 and then checked in 1977 with excellent results.
Further tests were performed on the USS Stanley and more recently on
the USS Cushing, increasing the support data for the corrosion
specification.

The corrosion program has since been expanded by NAVSEA to protect not only valves, but three general categories as listed in the new Specification DOD-STD-2138, "Metal Sprayed Coating Systems for Corrosion Protection Aboard Naval Ships". (USS Cushing test ship)

Category I -- Machining Space Components

1. Low pressure piping (7 to 10 mils thick).

2. Steam valves, piping and traps.

3. Auxiliary exhaust (such as stacks, mufflers and manifolds).

4. Air ejection valves.

5. Turnstiles.

Category II -- Topside Weather Equipment

A. Aluminum or Zinc Coating (7 to 10 mils thick)

1. Aircraft and cargo tie downs.

2. Aluminum helicopter decks.

3. Stanchions.

4. Scupper brackets.

5. Deck machinery coatings and foundations.

6. Chocks, bitts and cleats.

7. Pipe hangers.

8. Capstans.

9. Riggings and fittings.

10. Fire station hardware.

11. Lighting fixtures and brackets.

Category III

A. Aluminum or Zinc Coating (7 to 10 mils thick)

1. Decks in wash rooms and water closets.

2. Pump room decks and equipment support foundations.

3. Fan room decks and equipment support foundations.

4. Water heater room decks and equipment support foundations.

5. Air conditioning room decks and equipment support foundations.

6. Deck plate supports.

7. Machinery foundations.

8. Boiler air casings (Skirts).

This method of preservation is now being implemented throughout the Navy through its I.M.A., S.I.M.A. and shipyards. It has minimized the need for chipping and painting on a regular basis, and therefore released personnel for more skillful and technical duties.

NAVSEA is now in the process of generating corrosion control manuals for each class of ship; DD963, FFG, and CVN.

The process has also been introduced to the Army and is now in test aboard one of the landing craft at Fort Eustis, Virginia. We feel confident that the Army will utilize, as well as add to, the data base already generated by the Navy.

MEASUREMENT OF CAVITATION RESISTANCE OF
ORGANIC MARINE COATINGS

S. Basu and A.M. Sinnar
The Scientex Corporation
Washington, D.C. 20003

G.S. Bohlander
Coatings Application Branch
Nonmetallic Division
David Taylor Naval Ship R&D Center
Annapolis, Maryland 21401

ABSTRACT

As part of an overall U.S. Navy program to develop organic coatings for high performance craft so that they can better withstand harsh marine environments, eight different coatings from three generic classes (epoxy, polyurethane and elastomer) were investigated for their resistance to cavitation caused by hydro-dynamic impact. This paper describes in detail the development of a new measurement technique for cavitation resistance using a combination of the accelerated cavitation test method (ASTM G32-72 modified to test coated plates) and an acoustic emission monitoring system. The new technique is based on theoretical models which correlate acoustic emission signal properties to measurable cavitation parameters. The experimental procedure which involves the monitoring of acoustic signals generated during cavitation is described. Because of the nature of cavitation phenomenon, the signals are composed necessarily of cavitation source noise and acoustic emission. A method of approximating the acoustic emission content of the signals is discussed and an estimate of cavitation damage is computed using analytical relations. The results indicate that the new measurement technique is promising for the accelerated cavitation screening of organic coatings in marine environments provided certain problems in signal and noise analysis are overcome. Conclusions are drawn regarding the applicability and limitations of the proposed screening technique, and recommendations are made for future research in this area.

KEY WORDS

Acoustic emission; Analytical models; Cavitation damage; Cavitation resistance; Epoxy; Mean depth of penetration rate (MDPR); Organic coatings; Polyurethane; Screening technique.

INTRODUCTION

Cavitation erosion is a major problem encountered by high performance craft and high-speed hydrodynamic systems when they operate in harsh marine environments. To protect these systems against cavitation damage, both inorganic and organic coatings are customarily applied. The development of marine coatings that can better withstand erosive environments requires a laboratory technique which can evaluate the comparative cavitation resistance of various materials.

The cavitation resistance of any coating material is measured at present by the Standard Cavitation Erosion Test Method (ASTM G32-72)[1] modified to test coated plates.[2,3] In this method, cavitation bubbles are generated at high frequency and subsequently collapse on the face of a test specimen placed in the test liquid under specified conditions. The test is interrupted periodically and the specimen is removed, cleaned, dried and weighed to determine the loss of mass due to cavitation erosion. Test results are normally expressed in terms of cumulative mean depth of penetration (MDP) or cumulative weight loss versus test time. It is understood that a high weight loss or MDP value corresponds to low cavitation resistance and vice versa.

The screening method described above has certain limitations. First, it is not capable of continuously monitoring cavitation damage because the materials being tested tend to relax during the periodic interruptions in the testing process. Second, the analytical models used to correlate test results are based on the notion that all materials sustain volume or weight losses at a uniform rate. This theory is valid only for ideal homogeneous materials. In reality, all materials are inhomogeneous to some extent. Any estimate of the cavitation resistance of coatings which is based on assumptions of homogeneity is, therefore, likely to be inaccurate. This is particularly true for organic coatings. Consequently, the standard screening method has an unreliable predictive capability.

In a recent investigation,[4] the mechanics of cavitation was examined from the standpoint of the generation and propagation of pressure waves. It was postulated that damage to the solid surface caused by liquid particle impact manifested in the form of acoustic emissions. Based on this postulate, analytical models were derived.[5] These indicated the possibility of developing an improved measurement technique whereby the extent of damage due to cavitation could be related to measurable acoustic emission parameters. It is believed that this technique can not only determine more accurately the cavitation of marine coatings but also ultimately provide a more reliable tool for developing improved coating materials.

ANALYTICAL MODEL

Consider that χ represents a generalized cavitation damage variable. In a particular application, χ may denote volume loss,[6] mass loss[7] or even the mean depth of penetration.[8] The value of χ changes in time due to repeated impact loading. The process is similar to fatigue and, therefore, can be expressed mathematically by a growth rate equation. On the other hand, consider that μ represents a measure of acoustic emission activity. This measure is related to the amount of energy associated with the travelling pressure waves. The latter, in turn, is assumed to be related to the cavitation damage variable χ. Therefore, in theory, it is possible to obtain a functional relationship between the generalized damaged variable χ and the acoustic emission parameter μ as follows:

$$\chi = f(\mu) \tag{1}$$

The detailed derivation of specific forms of the above general functional relationship has been reported elsewhere by Basu.[5] Briefly, assuming that χ and μ represent the mean depth of penetration (MDP) and the root mean square value, V_{rms} of the signal voltage respectively, the following relationship was derived:

$$\chi = \left\{ \frac{(1-m\gamma)A_1 k^{-\gamma}}{(1-2\delta)A_2} \right\}^{\frac{1}{1-m\gamma}} \cdot V_{rms}^{\frac{1+2(\gamma-\delta)}{1-m\gamma}} \tag{2}$$

86

In the above expression, A_1 and A_2 are constants of the rate equations which govern the growth of χ and V_{rms} respectively, and γ and δ are the corresponding exponents of these equations. Also, m is another exponent in the growth rate equation of χ which incorporates the concept of a "cavitation intensity factor"[9] in a manner similar to that of the stress intensity factor in linear elastic fracture mechanics. Finally, in the above expression, k is a transmittance factor which relates V_{rms} to the amount of impact energy absorbed per unit volume, U_e, such that:

$$V^2_{rms} = kU_e \qquad , \qquad o < k < 1 \tag{3}$$

This is a reasonable assumption for it simply states that a fraction of the absorbed energy is transformed into acoustic emissions.

In theory, Equation (2) provides the necessary analytical tool to determine the degree of cavitation damage by continuously monitoring the acoustic emissions. However, it is frequently convenient to express the results of a cavitation screening experiment in the following forms:

$$\frac{dV_{rms}}{dn} = C_1 n^{q_1} \tag{4}$$

$$\chi = C_2 n^{q_1} \tag{5}$$

where C_1, C_2, q_1 and q_2 are estimated from the regression fit of experimental data. In this case, it is possible to relate these coefficients with those in Equation (2). The detailed derivation is given in reference (5) and will not be repeated here. Note that the term in the above expressions simply denotes the number of impacts.

EXPERIMENTAL PROGRAM

The main purpose of the experimental program was to determine the feasibility of using the acoustic emission technique as a quantitative tool for measuring the cavitation resistance of organic marine coatings. As such, the experiment consisted basically of measuring the acoustic emission signals for different coatings as functions of cavitation duration at simulated field intensities and under simulated hydrodynamic shear forces. The effects of coating thickness and substrate roughness were investigated to a limited extent. The effects of other parameters such as the curing temperature and time on the cavitation resistance of coatings were not addressed explicitly on this program.

Coatings and Substrates. Based on industry recommendations, field performance and laboratory test results,[3] eight coatings of three generic types were supplied by the David W. Taylor Naval Ship Research and Development Center (DTNSRDC), Annapolis Laboratory. These coatings and their chemical classifications (generic types) are listed in Table 1. Coatings were air sprayed to the appropriate thickness on 8 in. x 4 in. (203 mm x 102 mm) substrate panels cut from 1/4 in. (6.35 mm) HY-130 steel sheet stock, machined and ground to a surface finish roughness of 15 μin. (0.38 mm), and degreased in solvent. Coated specimens were allowed to dry for a period of 7 days or more at an average temperature of 70°F (21°C) and at 30% average relative humidity.

Simulation of Cavitation. The accelerated cavitation test facility at the DTNSRDC Annapolis Laboratory was selected for the simulation of cavitation

Table 1

Classification of Coating Materials

Coating Code	Chemical (Generic) Type
7133	Epoxy
PR475	Polyurethane
413	Polyurethane
LX500*	Polyurethane
LX500**	Polyurethane
Navy 150	Epoxy
M313	Polyurethane
Elastuff 504	Elastomer

* ceramic base   ** teflon base

intensity. The apparatus, shown in Figure 1[†] consists of a transducer fed by a 20kHz power supply which is acoustically coupled to an exponential titanium horn to produce controlled vibratory motions from 1.0 to 3.0 mils (0.025 mm to 0.075 mm) peak to peak at the working end of the horn. The test method conformed in general to ASTM G32-72[1] except that in this case, coated specimens were subjected to erosion by being placed directly under the titanium horn at a fixed distance of 20 mils (0.5 mm). In this manner, the cavitation intensity, as determined from the ratio of the maximum erosion rate sustained by the specimen to the maximum erosion rate at the horn tip (determined from the weight loss versus time plot), amounted to approximately 0.1 W/m^2. Note that this intensity corresponds to that encountered in service; however, the frequency of the laboratory test is much higher than that encountered in actual field conditions.

Acoustic Emission Instrumentation. The acoustic emission system consisted of a wide band transducer with a flat frequency response ranging from 100 kHz to 2 MHz, a 40/60 dB preamplifier with a selectable bandpass filter ranging from 10 kHz to 2 MHz, a high speed FFT spectrum analyzer and an X-Y plotter. The transducer was coupled to the preamplifier whose output was connected to the signal processing unit (FFT analyzer) and the X-Y plotter. The analysis range of the processor unit was between 10 Hz and 320 KHz full-scale with 95 selectable filters, a dynamic range greater than 65 dB and a noise floor of 70 dB below full scale. A frequency range extension up to 2 MHz was considered through an additional range extender unit. However, poor resolution prevented any meaningful signal analysis beyond 300 kHz range. The output from the signal analyzer was connected to an X-Y plotter. A photograph of the combined cavitation test facility and the acoustic emission instrumentation is shown in Figure 2.

Test Procedure. The basic test procedure consisted of placing a coated specimen under the cavitation horn and cavitating an area of the specimen. The acoustic sensor was mounted on the specimen at a distance of 2 in (50.8 mm) from the cavitation source. The specimen, along with the horn tip and the sensor, was kept submerged in a water container. A temperature control bath was used to maintain the water temperature at about 68°F (20°C). During cavitation, the signal from the sensor was amplified 40 dB by the preamplifier, processed in the frequency domain by the signal processor, attenuated 10 dB and displayed on the oscilloscope (or recorded in the plotter) in units of dB/1^V rms. The frequency range

---

[†]All figures appear at the end of the text.

of the plot was between 0 and 300 kHz. The 10 dB attenuation and in a few instances, 20 dB, was necessary to avoid possible overloading of the processor unit. The first analog record was taken at 15 seconds, the characteristic stabilization time of the signal. Thereafter, additional analog plots were made at 5 minute intervals until the end of the test. A typical frequency plot is shown in Figure 3.

The duration of tests varied from as little as 5 minutes, when slight cavitation damage was noticed in some specimens, to several hours for highly cavitation resistant coatings. Also, for each specimen, a series of tests with different test durations, was carried out at different locations in the specimen. This was achieved simply by moving or sliding the test specimen against the horizontal plexiglass support connected to the cavitation horn. The results from these tests provided a reasonably large data base which was later used to establish a quantitative relationship between the extent of cavitation damage and the level of acoustic emission.

RESULTS AND DISCUSSION

The raw data obtained from the experiment consisted of a large number of analog plots of signal rms values versus frequencies. It was noted that in the low frequency range (below 100 kHz), the output signal had a strong cavitation harmonics content. In contrast, the signal above 100 kHz and in particular, above 150 kHz, had a relatively small background cavitation noise (below 40 dB). Therefore, further analysis of signal characteristics was confined to the frequency range between 150 kHz and 300 kHz. Specifically, the acoustic emission content of the signals in this frequency range was estimated using the approximate analytical method described in reference (5). The method is illustrated graphically in Figure 4.

Figure 5 is a plot of the acoustic emission in millivolts from different coatings as a function of frequency. The figure clearly indicates the distinctly different acoustic emission characteristics of various coatings. If we consider that the acoustic emission is a direct representation of cavitation performance, it becomes evident how Figure 5 could be used as a preliminary and qualitative cavitation screening tool. In other words, the more the acoustic emission, the more severe the cavitation damage of coatings. On this basis, the coating type Navy 150 appears to have the least cavitation resistance with the sole exception of LX500 (ceramics). While polyurethane is considered to be a good coating material for marine applications and has reasonably good cavitation resistance, as evidenced by the acoustic emission characteristics of LX500 (teflon), the unusually low cavitation resistance of LX500 (ceramics) could be explained by its more brittle nature. Figure 5 also indicates that the coating types PR475 and 413 have excellent cavitation resistance. Two other materials, M313 and Elastuff 504, which do not appear in the figure, are also believed to have excellent cavitation resistance. However, this conclusion could not be corroborated here because of the lack of base data for M313 and Elastuff 504 in the high frequency range.

Figure 6 shows the rate of change of V_{rms} as a function of the number of impacts, n, for six different coatings. From the regression fit of experimental data using Equation (6), the regression coefficients C_1 and q_1 were evaluated for each coating.

Figures 7 and 8 show the micrographs (magnification 6.6x) of the cavitation damage in various coatings as a function of time. The damage, represented in terms of the average loss of coating thickness, was measured by a magnetic thickness measuring device. The mean depth of penetration was computed from these measuremens for coating types, 7133, Navy 150, LX500 (ceramics) and LX500 (teflon). Equation (5) was then used to fit the experimental data (see Figure 9) and subsequently, the regression coefficients C_2 and q_2 were evaluated. The values of C_1, C_2 and q_1 and q_2 are given in Table 2.

TABLE 2

Estimates of Regression Parameters for Various Coatings

Regression Parameters	Coating Types			
	7133	LX500 (ceramics)	LX500 (teflon)	Navy 150
C_1	2.08×10^{-5}	4.57×10^{-5}	1.84×10^{-5}	7.53×10^{-5}
q_1	−0.029	−0.035	−0.014	−0.081
r	−0.82	−0.88	−0.66	−0.58
C_2	1.33×10^{-12}	4.31×10^{-21}	7.20×10^{-19}	2.62×10^{-12}
q_2	1.01	2.16	1.95	1.04
r	0.98	0.99	0.98	0.98

Figure 9 may be used as a guideline for the cavitation screening of coatings. Note from the figure that Navy 150 coating has a different MDP rate when compared to coating 7133 although both are epoxy generic type. The difference is likely to be due to the addition of other chemical agents in the epoxy base which significantly change the cavitation resistance characteristics of these coatings. A similar observation can be made of the MDP rates of LX500 (ceramic) and LX500 (teflon) coatings. Here, the addition of ceramic ingredients to a polyurethane base in one case makes the coating more brittle and susceptible to cavitation whereas, in the other case, the addition of teflon ingredients to the same base makes the coating more resistant to cavitation.

Another guideline for the cavitation screening of coatings may be obtained by combining the information in Figures 6 and 9. Thus, Figure 10 shows the relationship between MDP and V_{rms} for various coatings. As indicated in the figure, the experimental results are reasonably in agreement with the analytical formulation (Equation (2)) represented by solid lines. To plot the latter, the values of A_1, A_2, γ and δ in Equation (2) were computed from those of C_1, C_2, q_1 and q_2 in the manner which is described in detail in reference (5). Also, the values of k and m were arbitrarily chosen as 0.5. A value of k equal to 0.5 simply means that fifty percent of the absorbed impact energy transforms into acoustic emission signals. This need not be the case. More appropriately, k should be considered as a function of the loss factor of coating materials conventionally expressed in terms of the ratio between the storage modules and the loss modules. Likewise, a value of m equal to 0.5 simply means that the cavitation intensity factor introduced earlier is a function of the square root of MDP.

CONCLUSION

The present method provides a quantitative estimate of the cavitation damage in coatings by continuously monitoring the output acoustic emission signals and

90

hence, effectively eliminates the need to interrupt the tests at periodic intervals. Therefore, it appears that the method can be used as a viable cavitation screening tool for organic coatings. For qualitative screening, one must have knowledge of the relevant material properties of coatings.

The analysis of acoustic emission signals reported in the text is an approximate one. The approximation leads to a more conservative estimate of the acoustic emission content of measured output signals and, correspondingly, the estimate of cavitation resistance of coatings. Such an estimate may be more desirable from an engineering standpoint. However, for greater accuracy and reliability, an improvement of the acoustic emission signal analysis may be necessary.

In view of the above conclusions, it is recommended that a major effort be directed toward developing an improved method of isolating acoustic emission signals from cavitation noise. Concurrently, an effort should be made toward identifying a common set of material properties that are correlated to the cavitation performance of organic coatings. Provided these efforts are undertaken, the acoustic emission technique holds excellent potential for the accelerated cavitation screening of coatings.

ACKNOWLEDGEMENTS

The research reported herein was partially funded by the David W. Taylor Naval Ship Research and Development Center (DTNSRDC), Annapolis, Maryland under Contract Number N00167-81-C-0281.

REFERENCES

1.   American Society for Testing and Materials, "Standard Method of Vibratory Cavitation Erosion Test," ASTM Designation G32-72, ASTM Standard, Part 31, July 1972, pp. 1179-1183.
2.   Hammitt, F.G., et al., "Liquid Impact Behavior of Various Nonmetallic Materials," ASTM STP 567, American Society for Testing and Materials, December 1974, pp. 197-218.
3.   Bohlander, G.S., and Preiser, H.S., "Characterization of Hydrofoil Coatings on the Basis of Adhesion Strength and Erosion Resistance," Report 76-0048, David W. Taylor Naval Ship Research and Development Center, Annapolis, Maryland, April 1976.
4.   Basu, S., "Preliminary Development of an Acoustic Emission Cavitation Screening Technique," The Scientex Corporation, Report No. TSC-16-2, September 1981.
5.   Basu, S., "Accelerated Cavitation Screening of Organic Coatings Using Acoustic Emission Technique," Proc. ASME Conf. Failure Prevention and Reliability, Dearborn, Michigan, September 1983 (also to appear in the Journal of Vibration, Acoustics, Stress and Reliability in Design).
6.   Lichtman, J.Z., et al., "Study of Corrosion and Cavitation-Erosion Damage," Transactions, American Society of Mechanical Engineers, Vol. 80, 1958, pp. 1325-1341.
7.   Leith, W.C., and Thompson, A.L., "Some Corrosion Effects in Accelerated Cavitation Damage," Transactions, American Society of Mechanical Engineers, Vol. 82D, 1960, pp. 795-807.
8.   Heymann, F.J., "On the Time Dependence of the Rate of Erosion Due to Impingement or Cavitation," ASTM STP 408, American Society for Testing and Materials, July 1966, pp. 70-100.
9.   Thiruvengadam, A., "The Concept of Erosion Strength," ASTM STP 408, American Society for Testing and Materials, July 1966, pp. 22-41.

Figure 1. Vibratory Cavitation Erosion Apparatus

Figure 2. Integrated Acoustic Emission Cavitation Screening
Set-up

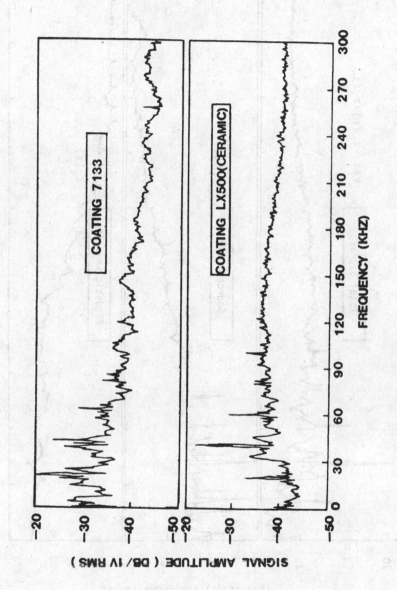

Figure 3. Typical Frequency Plot of Acoustic Emission Signals

93

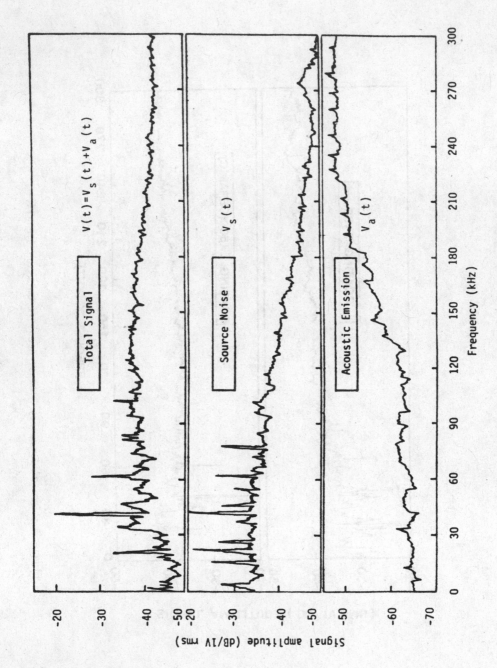

Figure 4. Representation of Acoustic Emission Signal Analysis in the Presence of Noise

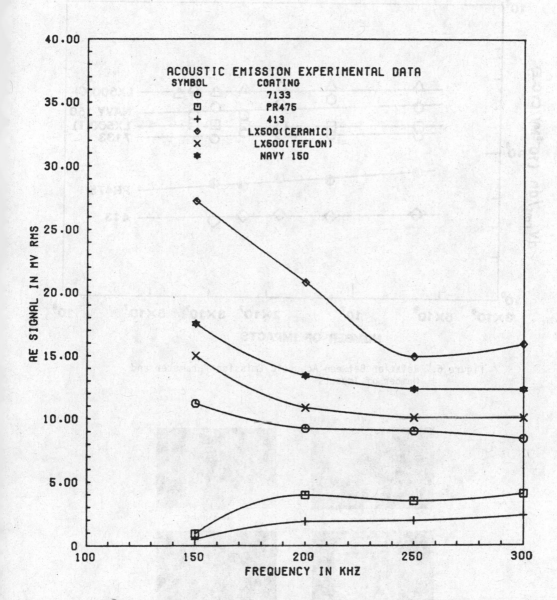

Figure 5. Acoustic Emission from Different Coatings
versus Frequency

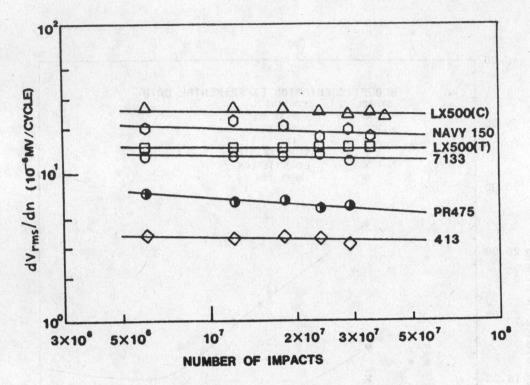

Figure 6. Relation Between Acoustic Emission Parameter and
Number of Impacts

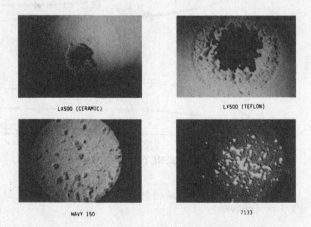

LX500 (CERAMIC)

LX500 (TEFLON)

NAVY 150

7133

Figure 7. Micrographs of Cavitation Damage of Various
Coatings (6.6 x before reduction)

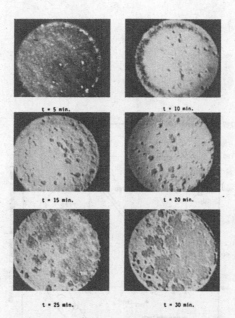

t = 5 min.   t = 10 min.

t = 15 min.   t = 20 min.

t = 25 min.   t = 30 min.

Figure 8.   Micrographs of Cavitation Damage Progression of Navy 150 Coating

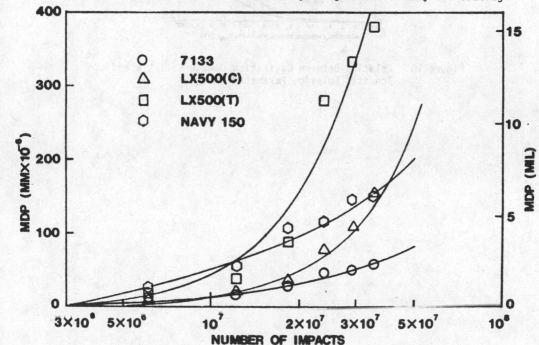

Figure 9.   Relation Between Cavitation Damage Variable and Number of Impacts

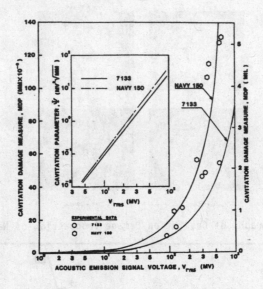

Figure 10.   Relation Between Cavitation Damage Variable and
Acoustic Emission Parameter

98

SESSION IV

FIELD APPLIED COATINGS FOR CORROSION CONTROL

CHAIRMAN: S. KETCHAM
           NAVAL AIR DEVELOPMENT CENTER

THE DISPLACEMENT OF WATER FROM A STEEL SURFACE

Charles R. Hegedus
Naval Air Development Center
Warminster, PA 18794

Abstract

A quantitative test has been developed to evaluate liquid compounds for their ability to displace water droplets from a steel surface and a study was made into the water displacement mechanism. The test method consists of the placement of water droplets onto an inclined steel surface, followed by the application of a test agent. The agent flows down the surface, contacting the water drops. The water may either remain on the specimen or be displaced. The specimen is then immersed in methanol which absorbs residual water on the specimen. The methanol is then analyzed for water content, yielding a quantitative result for water displacement.

Five materials have been evaluated. Two silicone alkyd compounds were found to be good water displacers. A third silicone alkyd compound was found to be a poor displacer at low angles but effective at higher angles. An acrylic and an epoxy coating were found to be poor water displacers.

ELASTOMERIC COATING SYSTEMS FOR NAVAL AIRCRAFT

D. Pulley
Naval Air Development Center
Warminster, PA 18794

Abstract

Tactical aircraft are subjected to considerable structural flexing during high-altitude maneuvers. This creates stresses within exterior coatings and causes these films to crack and chip. Saltwater can then penetrate to the metallic substrate, resulting in corrosion. Elastomeric coatings are made from long-chain polymers with a low crosslink density. Their rubbery nature allows them to dissipate flexing stresses, even at low temperatures.

The Navy is currently evaluating an elastomeric primer. It is a one-component, aliphatic polyurethane coating that cures with moisture from the air. Because of its unique combination of adhesion and flexibility, this coating can function as both a primer and a sealant at the same time. Another elastomeric coating is composed of a two-component, aromatic polyurethane binder reacted with a ketimine curing agent. This coating is able to dissipate stresses caused by the impact of rain and sand particles at high speeds. In addition to providing erosion protection, it has been used as a sealant over epoxy primer. Both of these materials have been used successfully on tactical aircraft.

EVALUATION OF OVER-RUST PRIMERS

S.J. Calabrese, F.F. Ling and S.F. Murray
Rensselaer Polytechnic Institute
Troy, New York 12181

Abstract: Several paint companies produce coatings which can be applied directly to rusted steel surfaces. An effective coating must perform the following functions: a) displace moisture at the steel surface, b) convert the existing rust to another compound of iron, or encapsulate the rust and bond it to the surface, and c) protect against further rust formation. A number of commercial coatings were identified which provided these functions. The results of an experimental program, in which some of these products were evaluated, are discussed in this paper.

Key words: Rust compatible primers; over-rust primers; marine coatings; hull coatings.

Introduction: There has long been a need for paint coatings or primers that can be applied directly over rusty metal surfaces. Surface preparation costs have escalated to the point where the cost for grit blasting to a "commercial" or "near white" finish is as much, or more, than the cost of applying the coating itself. Even if the cost is acceptable, there are occasions when it is extremely difficult to prepare the surface adequately. In view of the above, other methods of surface preparation have been considered. These include: A) weak acid cleaning (citric acid), Ref.[1]; B) absorption of the rust into an easily removable coating; and C) rust compatible primers.

Methods A and B are concerned with surface preparation and the removal of rust. In using rust compatible primers (C), the emphasis shifts to the concept of a coating protecting a prerusted surface from further corrosion with a minimum of preparation. This would normally mean that the loosely adherent rust should be removed first. The primer coat must then be applied to the surface and must create a bond between the adherent rust and the substrate. This is a "tall order" for a coating, but there are materials which show promise.

Many attempts have been made to combat rust by the use of a coating system [2, 3, 4]. Most of this work started in Europe, but recently, U.S. companies have been involved.

At this time, the most promising coatings are based on one of three principles [4]: rust stabilization; rust conversion; and rust penetration. While each has a different mechanism to inhibit rust, all coatings must have a second characteristic, which is the formation of a moisture or vapor barrier, to prevent further rusting. In addition, the primer must form a tightly adherent bond to the rust and the substrate.

There are various mechanisms by which the coatings appear to work [2]. Essentially, there are as follows:

> Stabilisation coating - converts the rust to a more stable iron oxide compound such as Fe_3O_4 (black). This is accomplished over a long period of time (9 to 12 months) after the coating has been applied.

> Conversion coating - reacts with the rust to form another compound of iron. The most widely used approach is the use of phosphoric acid, which converts the iron oxide to iron phosphate.

Penetration coating - penetrates the rust and protects the substrate from further corrosion. These coatings are usually based on a fish oil or linseed oil compounds.

There are problems involved with the use of these coatings. For instance, the stabilization coating and conversion coating must contain a large enough proportion of reactants to explicitly convert the rust. The penetration coating must be capable of complete penetration and, if needed, must be compatible with a topcoat. Still another variable is the presence of a solvent. Although the primer contains products to accomplish the rust arrestment, it is being used in a coating of some generic base resin (epoxy, polyurethane, alkyd, polyester, etc.) and most contain a carrier (water or solvent). Hence, the combination of rust combatment, resin, and solvent leaves the door open for a multitude of coatings.

In order to establish the effectiveness of these materials as viable coatings to inhibit rust, the U.S. Department of Commerce, Maritime Administration, funded a program with RPI through Avondale Shipyards to evaluate some of the coatings which are presently on the market. This paper discusses the results of that program.

Selection of Coatings: Inquiries were sent to several paint manufacturers to determine if they had a product which could be applied directly to a rusted surface. Those that responded positively were sent rusted steel coupon samples which were prepared in the laboratory at RPI. In some instances, the coating was applied at RPI when the supplier did not have an outlet for application.

A questionnaire was sent with the samples and filled out by the applicator. This essentially described the method of coating, the coating application conditions (temperature, relative humidity, etc.), type of coating, generic base, compatible topcoats, number and thickness of topcoats applied, and method of combating rust.

Coatings Evaluated: At least one primer coating was selected from each category. Table 1 shows the type of coatings evaluated and the topcoats, if any.

Note that the first three are mastic types, applied directly to the rusted steel. They are used as "high build" materials with no topcoat. The rest are primers and are applied with additional overcoats. In some cases, the topcoat was also a product of the primer manufacturer. Standard marine coatings were used where the primer manufacturer did not have a particular topcoat product.

TABLE 1 - RUST COMPATIBLE MATERIALS

Material	Type	Function	Thickness	Topcoat
Latex	Mastic	Stabilization	.040	None
Epoxy-water base	Mastic	Stabilization	.020	None
Epoxy-solvent base	Mastic	Stabilization	.014	None
Rust-converter latex	Sprayable coating	Conversion	.002	Various
Urethane	Sprayable coating	Stabilization	.002	Various
Vinyl	Sprayable coating	Conversion	.002	Various
Chlorinated rubber	Sprayable coating	Conversion	.002	Various
Fish oil penetrator, alkyd	Sprayable coating	Penetration	.002	Alkyd

Test Apparatus and Evaluation Procedures: Three types of exposure tests were used to determine the capability of the coatings to inhibit further corrosion. These were: pressure cycle, salt spray, and immersion test with scribe line.

The first evaluation was a pressure cycle test where a coated specimen was immersed in a pressure chamber filled with water. The specimen size was 2"x6"x1/4" thick. The pressure was raised to approximately 20 psi for ten minutes and then reduced to ambient. This procedure was repeated five times, and the specimens were examined. They were then placed back in the chamber and the above procedure repeated. At the end of 50 cycles the test was discontinued.

This test method is an excellent technique for evaluating coatings quickly. Those materials which failed early in the test were usually poorly bonded to the substrate. This procedure also simulated the conditions to which tank coatings would be exposed. The specimens were rated according to ASTM D714 Blister Rating Standard.

The second test was a salt spray evaluation in accordance with ASTM B117-73. Here, the specimens were placed in an environmental chamber and exposed to salt spray for 6000 hours. They were periodically examined and rated according to ASTM D714 Blister rating.

The third method was a simple immersion test where the specimens were scribed on a diagonal across the center of the specimen and allowed to soak in water containing 5% sodium chloride for several hundred hours. The depth of undercutting was measured and recorded. The specimens were rated according to ASTM Standard D1654 Method A.

The rusted surface samples were prepared at RPI by immersion in a fresh water tank, heated to 90°F. The water was continually agitated and aerated to produce an even layer of rust. The typical time required to produce the rusted samples was four weeks. Figure 1 shows the conditions of a rusted sample before coating.

Figure 1 Typical Surface Condition of a Rusted Sample
Before Application of a Candidate Coating

Test Results: The pressure cycle tests produced results which varied considerably, depending on the coating. These results are summarized in Table 2. The mastics gave the best results with essentially no change in the coating. The latex conversion primer + epoxy gave interesting results in that the surface looked excellent with no signs of blistering. However, after the test had been concluded and the specimens were allowed to set in the laboratory several hundred hours, the coating cracked and completely disbonded from the surface. Figure 2 shows the typical condition of some of the specimens after test. A comparison of the results indicated that the mastic coatings all showed little or no change, while all of the others, except for the latex conversion/epoxy system, had blisters. The blister size is inversely proportional to the number and the frequency is denoted by the density.

TABLE 2 - TABULATION OF PRESSURE CYCLE TEST

Primer/Topcoat	Thickness Primer/Topcoat	ASTM D714 Blister Rating 10 Cycles - 50 Cycles	Remarks
Latex mastic/none	.040"/none	None - None	Some rust bleedthrough
Epoxy water base mastic/none	.020"/none	None - None	Some rust bleedthrough
Epoxy solvent base mastic/none	.014"/none	None - None	Some rust bleedthrough
Urethane/urethane	.002"/.004"	#4 - 8 - #4 dense Medium dense	No rust through
Epoxy/alkyd	.002"/.004"	3 each #4 8 each #6 - Same	Some rust through
Latex conversion/ epoxy	.002"/.008"	None - None	Failed by disbondment after 500 hours
Latex conversion/ vinyl	.002"/.007"	#4 Medium #4 Medium	Same as above
Penetration primer/alkyd	.002"/.004"	#6 dense #6 dense	Poor bond between primer and topcoat

Salt Spray Tests: The salt spray test results are summarized on Table 3. In this instance, the solvented epoxy (mastic) gave the best results while the latex base and water based epoxy specimens both showed large blisters accompanied by edge deterioration. The urethane primer system also gave encouraging results with no blisters and very little rust on the edges after 6000 hours. The other coatings were removed at various stages of failure. Figure 3 shows the condition of the solvented epoxy mastic coating and the urethane coating as compared to some of the less successful candidates.

Immersion Tests: The immersion test results are shown in Table 4. The rating is in accordance with ASTM D1654, Creepage Standard (creepage is the amount of corrosion which takes place beneath the coating from the edge of the scribe). The range is from 00 (1 inch) to 10 (1/64 inch). The solvented epoxy mastic and urethane primer/ alkyd topcoat gave the best results. The water base and latex mastics showed edge deterioration at the scratch, while the latex conversion primers all gave higher creepage ratings accompanied by blisters. Figure 4 shows the condition of some of the specimens after test.

Discussion: The results indicate that the solvent-based epoxy mastic gave the best overall performance in these evaluations. The manufacturer indicated that this material was a combination of penetrator and stabilization coatings. After the tests were performed, a piece of coating was removed from the specimen to examine the condition of the rusted surface. The steel appeared clean, but the back of the coating contained black debris, probably Fe_3O_4.

The urethane, which also gave encouraging results, appeared to perform better with a nonurethane topcoat.

TABLE 3 - TABULATION OF SALT SPRAY TEST 6000 HOURS

Primer/Topcoat	Thickness Primer/Topcoat	ASTM D714 Blister Rating	Remarks
Latex mastic/none	.040/none	2 Medium	Edge deterioration
Epoxy water base mastic/none	.020/none	2 Few	Edge deterioration
Epoxy solvent base mastic/none	.014/none	None	Surface excellent
Urethane/urethane	.014/.005	None	Light rust through
Chlorinated rubber/ alkyd	.002/.004	Rust through	Discontinued after 500 hours
Latex conversion/ epoxy	.002/.002	4 Dense	Discontinued after 3000 hours
Penetration primer/ alkyd	.002/.004	1 Dense	Discontinued after 240 hours

TABLE 4 - IMMERSION TEST RESULTS

Material Combination Primer/Topcoat	Thickness in Inches Primer/Topcoat	ASTM D1654 Creepage Rating	Remarks
Latex mastic/none	.040/none	#8	Edge deterioration
Epoxy solvent base mastic/none	.014/none	#9	Excellent
Epoxy water base mastic/none	.020/none	#7	Edge deterioration, corner chipped
Urethane/urethane	.002/.004	#9	Some blisters - size #2
Urethane/alkyd	.002/.003	#9	Excellent
Latex conversion/ coal tar epoxy	.004/.008	#2	Poor
Latex conversion/ chlorinated rubber	.003/.002	#4	Poor
Penetration primer/ alkyd	.002/.004	#2	Coating removed between primer and topcoat

Conclusions: As a result of this evaluation, the following conclusions can be reached:
- The solvent-based epoxy mastic, which was both a penetrator and a rust stabilization type primer, gave the best results.
- The urethane stabilization primer, with an alkyd or epoxy topcoat, also gave encouraging results.
- The conversion type primers all suffered from continued rusting at the substrate, regardless of the topcoat system applied.

- The penetration type primer failed between the primer and the topcoat.

Recommendations: Additional testing should be performed on surfaces which have been rusted in an outdoor environment where exposure to industrial pollution as well as salt water has occurred since typical rusted surfaces can contain other combinations of compounds, besides Fe and oxygen.

Acknowledgement: The authors gratefully acknowledge Mr. John Peart of Avondale Shipyards, Inc. for his help and guidance in this program.

References:

1. Fultz, B.S., "Cleaning of Steel Assemblies and Shipboard Touch-up Using Citric Acid," U.S. Department of Commerce Maritime Administration, May 1980.
2. Dasgupta, D., and Ross, T.K., "Investigation of Rusty Steel for Painting," British Corrosion, 6, 6, 241-243, November 1971.
3. Guruviah, S. et al., "The Role of Rust Converters in Painting of Corroded Steel," Anticorrosion Methods and Materials, 5, 8-10, May 1980.
4. Anon., "A New Way to Fight Corrosion," Anticorrosion Methods and Materials, 22, 9-10, October 1975.

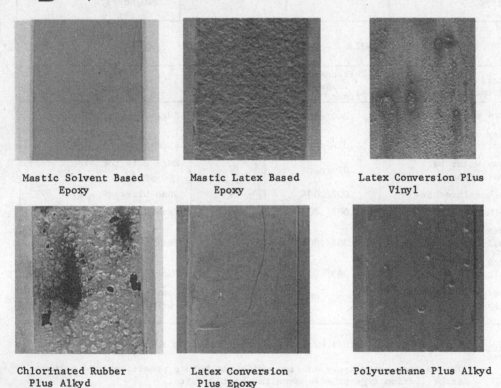

Mastic Solvent Based
Epoxy

Mastic Latex Based
Epoxy

Latex Conversion Plus
Vinyl

Chlorinated Rubber
Plus Alkyd

Latex Conversion
Plus Epoxy

Polyurethane Plus Alkyd

Figure 2   Typical Condition of Test Samples After Pressure
Cycle Test

Epoxy Mastic
Solvent Based

Epoxy Mastic
Water Base

Latex Conversion
Epoxy Topcoat

Polyurethane Primer
Polyurethane Topcoat

Figure 3 Condition of Specimen After Salt Spray Evaluation

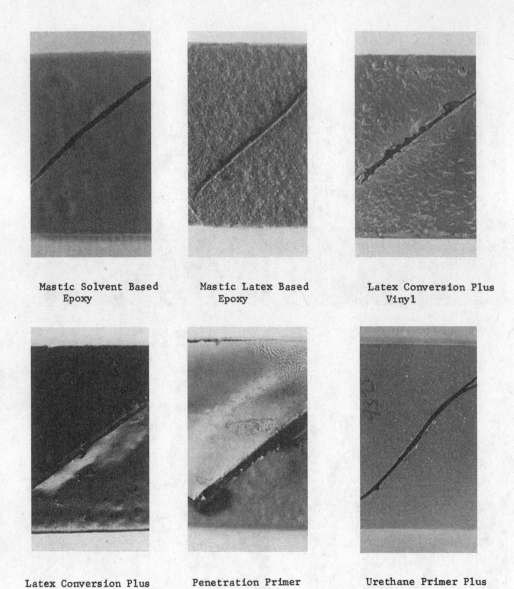

Mastic Solvent Based Epoxy	Mastic Latex Based Epoxy	Latex Conversion Plus Vinyl
Latex Conversion Plus Coal Tar Epoxy	Penetration Primer Plus Alkyd	Urethane Primer Plus Alkyd

Figure 4 Typical Condition of Test Samples After Exposure. 1500 Hours
Immersion Test

CORROSION PREVENTION FOR TACTICAL VEHICLES

M. J. Devine
General Technology
Havertown, PA 19083

D. V. Minuti
Naval Air Development Center
Warminister, PA 18974

M. B. Peterson
Wear Sciences
Arnold, MD 21012

Abstract: A common failure of motor vehicles and equipment constructed from steel is corrosion deterioration. The environmental exposure that promotes and accelerates this deterioration includes moisture, salt contamination and atmospheric pollution. Corrosion products and soils formed on the metal surfaces also hold water or other chemical compounds that may result in further metallic corrosion. The effects of this damage include (1) high repair costs, (2) shortened service life, (3) impaired safety, and (4) reduced availability of the vehicles or equipment. This paper identifies some of the existing technology advances in corrosion engineering that are readily applied and provide protection for metals exposed to diverse operational environments. The information presented covers selected materials and processes for the prevention and control of corrosion and the key elements for the implementation of a cost-effective program.

Key words: Abrasion; corrosion; corrosion prevention; erosion; fretting; protective coatings; tactical vehicles.

Introduction: The corrosion deterioration of motor vehicles has been the subject of a number of recent reports (1 - 5). The significance of the problem is frequently represented by data for the economic impact of corrosion, e.g., the Salt Institute reports that the annual cost of vehicle corrosion from road salting is $2 billion and an investigation by the National Bureau of Standards shows that annual costs of corrosion for personally-owned automobiles range from $6 to $14 billion.

Similar findings have been cited for tactical vehicles used by the military, e.g., the cost of the corrosion repair of 3800 motor vehicles and trailers was approximately $8 million. Corrosion deterioration can result in other adverse effects, viz., impaired safety, degraded reliability and reduced availability of vehicles.

Background: There are a number of critical factors that determine corrosion susceptibility for vehicles and equipment: e.g.:

   o Design
   o Manufacturing/Processing
   o Metal Treatment
   o Shipment/Storage conditions
   o Usage/Operating conditions

o Environment
o Type of Exposure
o Maintenance

If losses due to corrosion are to be avoided, a concentrated focus is required to assess the areas of opportunity and implement effective procedures during the various stages of the vehicle or equipment's life cycle.

This report is concerned with corrosion deterioration encountered during the operational or usage stage of the life cycle. A number of different vehicles including trucks, tanks and landing craft were examined in the field. A primary goal was to identify improved corrosion protection procedures to eliminate or minimize corrosion damage. The methodology consisted of:

o Assessment of Corrosion Failure
o Assessment of Improved Maintenance Procedures
o Application of New Technology
o Surveillance and Verification

In a recent study (6) concerned with corrosion and rust problems for tactical vehicles, it was stated that one reason for the problem was lack of effective maintenance. The emphasis on maintenance, especially preventive maintenance, cannot be overstated since detection and problem avoidance can be accomplished with the minimum expenditure of resources compared to the consequences of mechanical failure or the remedial measures necessitated by the lack of such maintenance programs. Preventive maintenance is the critical element since only a continuous cycle of inspection, cleaning and protective coating touch-up will drastically reduce initiation of corrosion from salt, contaminants and water. Where severe operating conditions exist and impact damage causes loss of corrosion protection, repair of the deterioration should be accomplished at the earliest possible time based on maintenance procedures defined for organizational-level application.

Analysis: The key issues involved in the corrosion assessment and application program are listed below:

1. Environmental Effects
2. Vehicle/Equipment Susceptibility to Corrosion
3. Configuration Factors
4. State-of-Technology
5. Corrosion Protection
6. Maintenance Concepts
7. Definition of Effective Materials and Procedures
8. Requirements for implementation of improved Corrosion
    Prevention and Control

The exposure of tactical vehicles to a wide variety of severe operational and environmental conditions represents a major factor in corrosion deterioration and reduced service life. The following are examples of the different types of operational and environmental exposure:

Humidity	Stresses
Salt Water	Sand
Pollution	Airborne Debris
Silt	Micro-Organisms

112

The examination of such vehicles showed extensive corrosion damage involving structural, mechanical and electrical system deterioration. In addition to environmental exposure, a variety of factors contribute to the vehicle corrosion damage, e.g.,

o Lack of drainage holes, (bottom of doors and vehicle body areas)
o Lack of uniform rust-proofing
o Failure of provide adequate protective coatings
o Incompatible combinations in metals
o Moisture absorbing materials in contact with metal surfaces
o Lack of sealant in crevices

Examples of vehicle components, parts and assemblies that were found highly susceptible to corrosion are as follows:

Frames	Doors
Fenders	Hinges
Light Assemblies	Fuel Filters
Battery Compartments	Cables
Electrical Connectors	Floors
Weld Areas	Fasteners
Brake Parts	Panels

A brief outline of the corrosion analysis for several of the above areas is presented below:

Doors: Corrosion is frequently found on the exterior of the door (top or bottom) and on the interior at the bottom (Figure 1). In this condition the door is usually replaced. Corrosion treatments had not been applied on the interior surfaces of the door during manufacture.

Panels: Corrosion proceeds from the interior to outer surfaces. The severity can be assessed from the corrosion damage shown in Figure 2.

Battery Box: Corrosion of the battery box and accessories (Figure 3) is accentuated by battery acid contamination. Tie down brackets and bolts also become corroded.

Frames: Corrosion is found on the underside of the bed and on the cross frame support (Figure 4). Such damage at weld points can cause failures since vibration will accelerate fracture.

Lights: Moisture condenses inside the light due to temperature changes. This causes corrosion of the socket which interrupts the circuit. In order to repair the light the front cover must be removed. The corrosion of steel screws mounted in the aluminum housing results in additional damage during removal. The damage is shown on Figure 5.

A comprehensive listing of corrosion prone areas for three different types of vehicle is presented in Figures 6 - 8.

Corrosion Protection Technology: The concepts of Reliability-Centered Maintenance wherein the maintenance procedure is directly related to failure mode and consequences of failure were the basis for defining appropriate corrosion protection measures. The initial assessment covering a range of vehicles was essential to defining failure modes and causes. The latter included: (1) fretting, (2) erosion, and (3) abrasion. Based on demonstrated technology for advanced coatings capable of providing protection under such conditions, suitable materials and treatments were defined for tactical vehicle applications.

Some of the factors evaluated in the selection of the materials for corrosion prevention and control are shown on Figure 9. The basic functions of these materials are outlined in Figure 10. The essential properties that were considered for the protective coatings are in Figure 11.

A brief description of several of the selected protective coatings, including applicable specification designation, is presented below:

Specification	Description
MIL-L-63460	Cleaner and Preservative for Weapons and Weapons Systems
MIL-S-81733	Sealing and Coating Compound, Corrosion Inhibitive
MIL-C-85054	Corrosion Preventive Compound, Water Displacing, Clear
MIL-C-81309	Corrosion Preventive Compound, Water Displacing, Ultra Thin Film
MIL-L-23398	Lubricant, Solid Film, Air Drying
MIL-G-81322	Grease, General Purpose
MIL-C-0083933	Corrosion Preventive Compound, Cold Application

The critical steps for the preparation of an effective coating system included:

o Cleaning
o Corrosion Removal
o Metal Surface Preparation
o Pretreatment
o Sealing
o Undercoat
o Topcoat
o Inspection

Methods were defined for each of the different coating materials and applications. A few of the specific applications and protective coatings are as follows:

Application	Coating
Light-internal	MIL-C-81309, Type 3
Light-threads	MIL-L-63460
Light-external	MIL-C-85054
Fuel Filter	Nylon 11
Battery box	Nylon 11
Rocker Panel	MIL-C-0083933

The protective coatings were applied to a number of corrosion prone components of tactical vehicles and subjected to field operations including salt water exposure.

Results: The results of field evaluation established that after an exposure time of one year the coating materials selected continued to provide effective protection from corrosion damage. It was observed that for each of the control parts, included for comparison, corrosion ranged from mild to extensive. It is also important to note that the time required to apply the protective treatments was approximately 2.5 hours per vehicle. Considerable cost avoidance is projected based on this approach in comparison to repair and replace maintenance as a result of corrosion.

It is concluded that the successful implementation of the improved corrosion prevention and control procedures must consider the following factors:

o Training
o Facilities
o Engineering
o Support

It is further concluded that a systems approach to corrosion prevention is vital for both planned acquisition and fielded vehicles. The systems approach requires the following elements:

o Corrosion Program Plan
o Corrosion Prevention Guide - Vehicle Design and Development
o Corrosion Maintenance Plan
o Corrosion Malfunction Data Base
o Technology Support

These elements, when combined, can accelerate the transfer of technology improvements to design and maintenance and achieve long term corrosion protection for tactical vehicles.

References

1. National Bureau of Standards Special Publication 511-1; "Economic Effects of Metallic Corrosion in the United States" (May 1978).

2. Regan, T. W.; "War on Rust", Army Logistician (July - August 1982).

3. Technical Report No. 226; "An Evaluation of the Rust Condition of Truck, ¼ Ton: M151 Series, U.S. Army Systems Analysis Activity (February 1978).

4. Technical Report No. 291; "Survey of Rust Damage to the M880 Series, 1-¼ Ton Truck", U.S. Army Materiel Systems Analysis Activity (January 1980).

5. "Benefits and Costs of Road Salting"; (A Summary of a Study by the Institute for Safety Analysis) Reprinted by The Salt Institute (January 1981).

6. "Corrosion and Rust Problems for Army Vehicles"; Government Industry Data Exchange Program (GIDEP), Tape E246 - 1069 (March 1981).

(a) Outside

(b) Inside

Figure 1. Door Corrosion

(a) Severe Corrosion

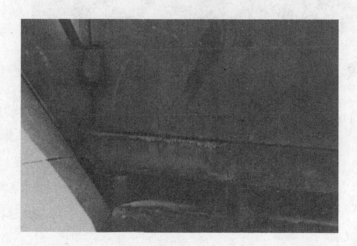

(b) Repair

Figure 2. Rocker Panel Corrosion

Figure 3. Photograph of Battery Box Components

Figure 4. Bed and Crossframe Corrosion

(a) Front Running Light

(b) Stop Light

Figure 5. Photographs of M-54 Truck Lights

Amphibious Vehicles (Example: LVT 7 Series)

o Battery Box/Clamps

o Oil Line Tube Joints

o Plenum Actuating Cylinder

o Fan Drive Arm

o Generator Turnbuckle

o Turret Cannon Plug

o Ammunition Case Hinges

o Turret Slip Ring Assembly

o Transmission Solenoids

o PTC Handle

o Final Drive Dip Stick

o Push-Pull Control

o Transmission Housing

o Bell Crank

o Seat Guide

o Bilge Pump

o Driver Ramp Control

o Horn

o Plenum

o Infrared Hatch

o Hydraulic Line Tube Joints

o Brake Arm Assembly

o Hydraulic Cylinder

o Headlight

o Tail Light

Figure 6. Specific Corrosion Problem Areas

M-60A1 Tank

o Powerplant Mounts, Guides and Brackets

o Fuel Tank Filler

o Fuel Tank Emergency Filler

o Fuel Tank Shut-off Valve Cable and Related Parts

o Accelerator Controls - Engine Compartment Linkage Assembly

o Electrical Connectors

o Tail Light - Stoplight Assembly

o Battery Box

o Engine Compartment

o Brake Actuating Lever and Moving Parts

Figure 7. Specific Corrosion Problem Areas

121

M Series (Wheeled Vehicles)

o Frame/Cross Members	o Windshield Frame
o Front Floor	o Mirror Attachment
o Rear Floor	o Threaded Fasteners
o Front Fenders	o Brake/Accelerator Brackets
o Rear Fenders	o Antenna
o Hood/Body	o Cables
o Tailgate	o Fuel Filters
o Rain Gutter	o Light Assemblies
o Doors	o Hinges
o Panels	o Battery Frame/Holder

Figure 8. Specific Problem Areas - Marine Corps Vehicles/Equipment

Factors

o Protection Properties
o Stability
o Ease-of-Application
o Ease-of-Reapplication
o Availability; e.g., Approved Supply System Materials (NSN)

Figure 9. Selection of Materials for Corrosion Prevention/Control

o To Remove Salt, Soils and Contaminants

o To Displace Water

o To Prevent Entry of Moisture in Crevices

o To Protect Metal Surfaces from Atmospheric or Corrosive Substances

Figure 10. Basic Functions

o Water Resistance

o Low Water Absorption

o Flexibility

o Uniformity

o Chemical Resistance

o Weather Resistance

o Abrasion Resistance

, o Ease of Application

o Adherence

o Appearance

Figure 11. Key Properties - Protective Coatings

CORROSION TESTS OF ELECTRODEPOSITED COATINGS
IN BOILING 100 PERCENT PHOSPHORIC ACID

C. E. Johnson*, J. L. Mullen**, and D. S. Lashmore*
Electrodeposition Group*
Corrosion Group**
Metallurgy Division
Center for Materials Science
National Bureau of Standards
Washington, DC 20234

Abstract

This paper reports the results of a feasibility study that is being carried out
on the use of electrodeposited metallic coatings for the protection of copper
heat exhangers in phosphoric acid based fuel cells. Electrodeposited coatings
of gold, silver, ruthenium, lead, metallic glass alloys of nickel-phosphorus and
cobalt-phosphorus, and duplex coatings of gold:lead and silver:lead on copper
substrates were evaluated for corrosion resistance in 100 v/o phosphoric acid at
200°C by weight-loss measurements and atomic absorption analysis of copper in
solution. To date, corrosion rates of DC electrodeposited coatings have been
determined by weight-loss measurements and extrapolated to mm/yr. Coating porosity,
cracking, and penetration were monitored by atomic absorption analysis of copper
detected in solution, reported in ppm. The duration of the tests for weight-loss
determination was 1000 hours. The metallic glass alloys did not provide corrosion
protection beyond 200 hours of immersion in 100 v/o phosphoric acid at 200°C.
Silver was the only coating tested that provided corrosion protection for 1000
hours. The remaining coatings fell in a range of 200 to 750 hours.

Key Words: Corrosion resistance; electrochemical fuel cell; electrodeposited
metallic coatings; metallic glass alloys; phosphoric acid.

Introduction

The corrosion resistance of electrodeposited metallic coatings to concentrated
phosphoric acid at a temperature of 200°C was investigated. This investigation
resulted from serious metallic corrosion of copper and aluminum components of
pertinent cooling and heat exchanger sub-systems in the Department of Energy
sponsored phosphoric acid based fuel cell program. The available literature
was reviewed and corrosion data in phosphoric acid at elevated temperatures
of metals suitable for electrodeposition are given in Table I (1,2,3,4).
Also included in this table are several non-metals reported just for comparison.
Although not all of these materials can be deposited from aqueous electrolytes,
they can be electrodeposited from either aqueous or molten salt electrolytes.
Even though the corrosion data reported in Table I is for solid materials and

not coatings, it still can provide a basis for selection of possible electrodeposited coatings from aqueous electrolytes, such as the underlined materials
and their alloys.

Experimental Procedure

Electrodeposited coatings were prepared on cylindrical copper substrates,
4.75 mm in diameter by 8.25 cm in length. This diameter rod was chosen to
simulate the actual size of the copper heat exhanger tubes used in the fuel
cells. The substrates were plated as rotating cyclindrical cathodes at 1000 rpm.
The electrolytes used for electrodeposition were prepared in the laboratory
according to standard formulations except for the 24 K gold and ruthenium which
were commercially available proprietary electrolytes. The compositions of the
various electrolytes will not be described, but the various type of coatings
that were investigated are listed in Table II. The coating deposition
techniques used were electroless (autocatalytic), direct current (DC), and
pulsed current.

Initially, electrodeposited coatings of metallic glass alloys were investigated
because of their good corrosion resistance as a result of their amorphous
structure. The reported corrosion resistance for these materials resulted
from corrosion tests in dilute hydrochloric and sulfuric acid solutions at
ambient temperatures. Copper substrates electrodeposited with noble metals
such as lead, gold, ruthenium, and silver were also prepared for evaluation.

The degree of corrosion protection provided by the coatings was evaluated in
cooperation with the corrosion group of the National Bureau of Standards
(NBS), using weight-loss measurements to obtain corrosion rates reported in
mm/yr. Atomic absorption analysis of the phosphoric acid was done simultaneously
with the weight-loss measurements to detect the presence of substrate metal
ions during exposure at 200°C. The detection of the copper substrate ions in
the phosphoric acid gave indication of porosity, cracking, or penetration of
the metallic coatings after prolonged exposure. Small aliquots of phosphoric
acid were withdrawn periodically during the weight-loss corrosion tests for
the atomic absorption analysis. The weight-loss corrosion tests were made in
a cell schematically illustrated in Figure 1. The cell was sealed to external
air except through the reflux chamber, thus the oxygen content in the phosphoric
acid was low, although the actual content was not measured. The coated
sample was only partially immersed in the phosphoric acid to determine the
corrosion effect of total immersion, as well as, an area that may be subjected
to phosphoric acid mist or spray.

On completion of a particular corrosion test, the sample was removed and
rinsed with distilled water and then ultrasonically cleaned in warm distilled
water to remove any loose scale. The samples were dried to a constant weight
and were weighed on an analytical balance accurate to 0.5 mg.

The corrosion (or dissolution) rate was calculated in mm/year using the
following formula:

$$x = \frac{(\Delta W) \ (8.76 \times 10^6)}{(\rho) \ \ (A) \ \ \ (T)}$$

This formula was derived as follows:

$$p = \frac{(\Delta W)(10^3)}{(\rho)(A)}$$

$$h = (t)(1.142 \times 10^{-4})$$

$$x = \frac{p}{h}$$

where p is corrosion (mm), ρ density (g/cm^3), A area (mm^2), t test time (hours), h test time (years), Wi initial weight (g), Wf final weight (g), ΔW (Wi-Wf), and x corrosion rate (mm/yr). Theoretical density values were used in the calculations for the single element noble metal deposits and measured values for the metallic glass alloys (5).

Results and Discussion

Table III and Figures 2 and 3 give the corrosion/dissolution data determined in this work, as well as, the data from the atomic absorption analysis of the phosphoric acid for copper metal ions from the substrate.

The results of the weight-loss corrosion tests of the metallic glass alloys, Table III and Figure 2, indicate that all of these coatings have a high corrosion rate in 100 v/o H$_3$PO$_4$ at 200°C Most of these coating had varying degrees of cracking after the corrosion testing. An indication of this behavior is revealed by the varying amounts of copper, from the substrate, detected in the phosphoric acid by atomic absorption analysis. The lowest corrosion rate for a metallic glass alloy was found for a coating from an electroless nickel electrolyte, but the amount of copper detected in the phosphoric acid was significantly higher indicating severe cracking of the coating, whereas, an electrodeposited metallic glass alloy coating 14 percent phosphorus from an electrolyte containing saccharin as a stress reducer produced a lower concentration of copper in the phosphoric acid but a much higher rate of corrosion. This implies that the cracking is resulting from stress in the coating which is reduced by the addition of saccharin to the electrolyte, but sulfur from the saccharin is being incorporated in the coating resulting in a higher corrosion rate. Of the metallic glass alloys, the cobalt - 10% phosphorus coating resulted in the highest corrosion rate.

The results of the corrosion tests of the electrodeposited noble metal coatings are shown in Table III and Figure 3. Electrodeposited silver was found to be far superior in corrosion resistance and lack of porosity or cracking, indicated by the low ppm of copper detected in the phosphoric acid after exposure of over 1700 hours, than any coating that has been investigated. The corrosion resistance of the ruthenium coating was good but the coating was severely cracked after testing resulting in significant amounts of copper detected in the the phosphoric acid. Presently there are three commercially available electrolytes for ruthenium plating but all are limited to crack-free deposits only of thicknesses less than 2.5 μm. The duplex coatings of silver over lead and gold over lead performed marginally well, indicating that this technique should be explored further. A duplex coating scheme would be to apply one coating that would be preferentially corroded with respect to an undercoating.

126

The electrodeposited 24 K gold deposit performed well for approximately 600 hours but then catastrophically failed, Figure 3, due to dissolution and porosity of the coating. The electrodeposited lead coating appears to have the same behavior as the gold, Figure 3.

This work is being continued with electrochemical corrosion measurements of the coatings and these results will be validated by the weight-loss data. If this validation is accomplished, the electrochemical corrosion technique will be used as an accelerated test. A teflon cell is being modified for the 200°C electrochemical corrosion measurements. These measurements will be made at the open circuit potential for the various coatings, as well as, at a positive 600 mv referenced to hydrogen.

Conclusions

1. Corrosion and/or dissolution rates were determined for several coatings, in concentrated phosphoric acid at 200°C for the first time.

2. Results of weight-loss measurements have shown that electrodeposited silver offers excellent protection to copper substrates in this severe environment.

3. Electrodeposited ruthenium was found to have low dissolution rates but the protection offered was poor due to cracking of the deposit.

4. Electrodeposited gold and lead performed well for approximately 600 hours then failed due to dissolution and porosity.

5. The metallic glass alloys performed poorly in this severe environment.

6. Duplex, ternary, and sacrificial coatings may be useful coatings, but further work is needed to confirm the preliminary results.

References

1. L. Marcus and R. Ahrens, "Chemical Resistance of Solid Materials to Concentrated Phosphoric Acid", Ceramic Bulletin, Vol. 60, No. 4, pp 490-493, April 1981.

2. F. L. LaQue and H. R. Copson, Corrosion Resistance of Metals and Alloys, 2d ed., pp 294, 632, 637, 642, Edited by F. L. LaQue and H. R. Copson, Reinhold, New York, 1963.

3. C. J. Smithells and E. A. Brondes, Metals Reference Book, 5th ed. p 147, Edited by C. J. Smithells, Butterworths (Plenum), New York, 1976.

4. I. Mellan, Corrosion Resistant Materials Handbook, 3d ed., p 210, Edited by I. Mellan, Noyes Data Corporation, Park Ridge, N. J., 1976.

5. A. Brenner, D. Couch, and E. Williams, "Electrodepositon of Alloys of Phosphorus with Nickel or Cobalt," J. Res. Natl. Bur. Stds., Vol 44, pp 109-122, Jan. 1950.

TABLE I

CHEMICAL RESISTANCE OF METALS IN H_3PO_4

	TEST CONDITIONS		CORROSION RATE
MATERIAL	WT%H_3PO_4	TEMP, °C	MM/YEAR
TANTALUM	85	191	0.005
TANTALUM	85	146-210	EXCELLENT
PLATINUM	88	204	<0.13
NIOBIUM	85	100	0.084
NIOBIUM	88	204	<0.13
NIOBIUM	20	85	POOR
TUNGSTEN	96	204	0.061
MOLYBDENUM	96	204	0.008
GOLD	88	204	<0.13
SILVER	88	204	<0.13
SILVER	85	79	1.04
COPPER	76	16	0.20
LEAD	76	21	1.45
NICKEL	90	85	3.81
TITANIUM	10	100	8-10
TITANIUM	20	85	POOR
ZIRCONIUM	85	160	51
VANADIUM	85	160	DISSOLVED
GRAPHITE	96-100	204	-16.61-3.00
VITON (191)	96	204	-0.008
TEFLON (TFE)	96	204	0.018
KEL-F 81	96	204	0.031

TABLE II

Electrodeposited Coatings Investigated

Metallic Glass Alloys:

Coating	Thickness, μm	Comments
Ni-12P	175	
Ni-13P	138	
Ni-14P	112	
Ni-14P	38	
Ni-16P	38	
Ni-14P	150	Saccharin added to reduce internal stress in deposit
Ni-14P	162	Saccharin plus stress relief at 150°C for 16 hrs.
Ni-10.5P-Mo	38	Sodium molybdate added to Ni-P solution.
Ni-10.5P-Mo	75	" " " " "
Ni-12P-Mo	50	" " " " "
Ni-12P-Mo	90	" " " " "
Ni-14P-Mo	25	" " " " "
Co-6P	225	
Co-6.5P	162	
Co-6.5P	200	
Co-10P	112	
Co-6P	338	Saccharin plus sodium lauryl sulfate as anti-pitting agent.
Co-6.5P	238	" " " " " " "
Co-6.5P	212	Saccharin plus FC-95 surfactant as anti-pitting agent.
Co-8P	200	Saccharin plus sodium lauryl sulfate as anti-pitting agent.
Ni-8P	262	Electrodeposit from an Electroless Nickel type solution.
Ni-9.5P	95	" " " " " " "
Ni-10.5P	30	" " " " " " "
Ni-8.5P	62	Electroless Nickel
Ni-9P	62	" "
Ni-10.5P	13	" "

Other Metals:

Coating	Thickness, μm	Comments
24K Au	13	
24K Au	50	
24K Au	50	
Ag	100	
Ag	138	
Ag	175	
Pb	150	
Pb	150	
Pb	165	
Ru	2.5	Undercoat of 24K Au 2.5 μm thick.
Ru	7.5	" " " " "
Au-Pb	Duplex	7.5μm 24 K Au over 62μm Pb.
Ag-Pb	Duplex	15μm Ag over 62μm Pb.

TABLE III

CHEMICAL RESISTANCE OF ELECTRODEPOSITED METALS
IN 85% H_3PO_4 AT 200°C

Metal Coating	Soak Time HRS	A.A. Copper Analysis PPM	Corrosion Rate MM/YEAR
Silver	1704	0.84	0.0045
Ruthenium	168	46.80	0.026
Silver : Lead	960	3.53	0.057
Gold : Lead	456	4.60	0.084
Gold	744	313.75	0.470
Lead	504	2.85	0.740
METALLIC GLASS ALLOYS:			
EN (Ni - 9P)	240	38.50	1.69
Ni - 14P	192	19.00	3.10
Ni - 8P (EN Type)	216	29.00	3.98
Ni - 12P	45	4.90	4.85
Ni - 16P	22	4.00	5.04
Ni - 12P - Mo over EN	96	10.96	6.64
Ni - 12P - Mo	72	---	7.00
Ni - 14P + Saccharin	50	1.80	7.84
Co - 10P	48	18.63	31.46

CORROSION CELL FOR WEIGHT-LOSS MEASUREMENTS

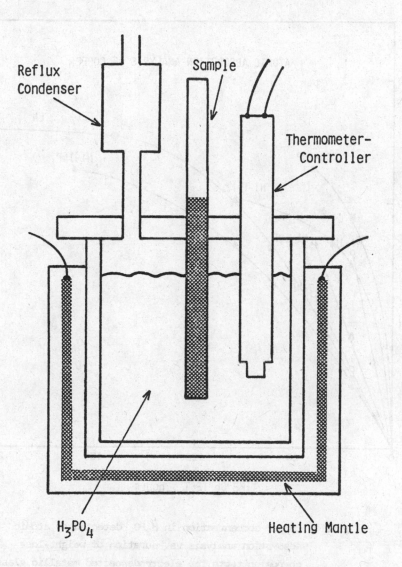

Figure 1. Schematic drawing of corrosion cell for
weight-loss measurements.

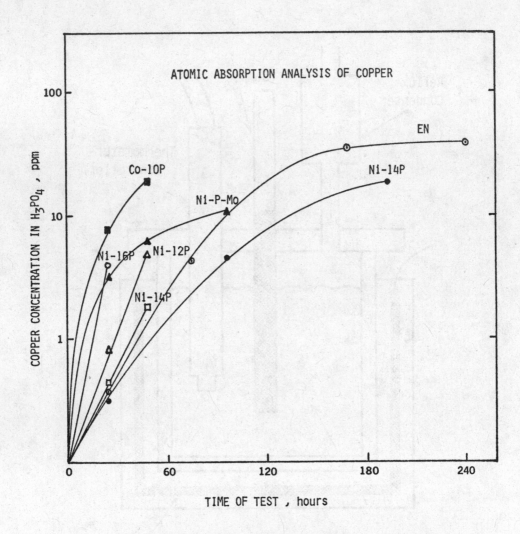

Figure 2. Copper concentration in H_3PO_4 detected by atomic absorption analysis vs. duration of weight-loss corrosion tests for electrodeposited metallic glass alloys.

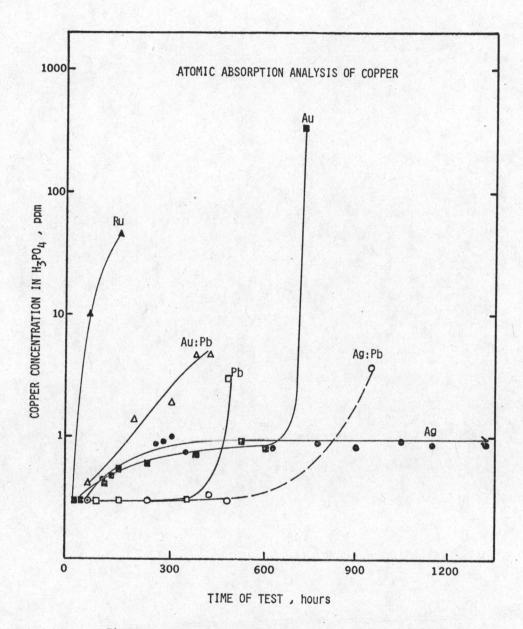

Figure 3. Copper concentration in H_3PO_4 detected by atomic absorption analysis vs. duration of weight-loss corrosion tests for electrodeposited noble metals.

SESSION V

TRIBOLOGICAL COATINGS

CHAIRMAN: P. J. BLAU
           NATIONAL BUREAU OF STANDARDS

A STRATEGY FOR SELECTION OF TRIBOLOGICAL COATINGS

M. B. Peterson
Wear Sciences, Inc.
Arnold, MD 21012

A. W. Ruff
Metallurgy Division
National Bureau of Standards
Washington, DC 20234

Abstract   Tribological Coatings are becoming increasingly important in the design
of mechanical equipment. The proper coating correctly applied can result in first
cost reductions, maintenance cost reductions, materials cost reductions, materials
and energy conservation, and significant improvements in performance. A wide
variety of tribological coatings and application techniques are available for use.
The primary problem that exists is in knowing what coating to select or develop for
any given application and what tests to conduct to evaluate coating performance.
A strategy for the selection of a tribological coating is proposed in this paper.
The important factor to consider is the tribological function of the coating. The
most important functions are reduction of various modes of wear, retention of
lubricant in the contact area, increase of lubricant load capacity, replacement or
rebuilding of contact surfaces, and modification of the coefficient of friction.
Once the required function needed in an application has been defined it is then
possible to select compositions and properties which will accentuate each function.
Specific laboratory tests can also be performed which rank the coatings as to
their ability to perform each function. Examples are given here for important
tribological functions. Over the past 10 years many new developments have been
made in metallic coatings to improve their utilization for tribological purposes.
These include composite and alloy coatings, new application techniques, simple
cost effective application techniques, and precise composition and microstructure
control. Examples are given here of advances in some of these areas.

Key words: Adhesion; alloys; coatings; friction; surfaces; tribology; wear.

I. Introduction   The use of coatings in tribological applications is growing
and will continue to grow as evidenced by a literature survey we have conducted
going back to 1965. Coatings can offer certain economic and technical advantages over the use of uncoated materials. Their main advantage is that coatings
allow the base material to be optimized for strength purposes while the
surface is optimized for wear, corrosion resistance, appearance, or thermal
behavior, for example. Furthermore, replacing the coating during rework or repair
may be more cost effective than new part manufacture. The principal disadvantage
in using coatings concerns the possibility of separation from the substrate during
use. While the discussion to follow emphasizes the considerations important in the
selection of tribological coatings, it should be noted that other alternatives may
exist for any particular problem. This could involve, for example, use of a more
effective lubricant, or a redesign of the mechanical system elements.

Tribological coatings are used for three different purposes:

1. Replace surfaces (e.g., Hard surfacing)

2. Modify surfaces (Surface alloying)
  3. Lubricate surfaces (e.g., Solid lubricant coatings)

Resurfacing essentially replaces one surface with another having presumably more desirable friction and wear properties. Usually the new surface is harder than the surface replaced but not always. Surface treatments modify the surface by hardening or by changing the composition or structure. The modified surface has unique tribological properties relative to the base metal. Solid lubricant coatings are soft and prevent surface damage by shearing more easily than the materials on which they are applied.

A wide variety of coating compositions is available. Table I lists those most frequently used for tribological purposes, based on our review of the literature. As can be seen they cover a wide variety of properties and compositions from very hard carbides to soft metal coatings. Each of these compositions can be applied by a variety of techniques listed in Table II. Some of these application techniques are very simple and inexpensive such as painting. Others are very complex either requiring vacuum processing or requiring a series of treatments and pretreatments. Some can be applied in the field while others can only be applied at particular facilties. Thus there is no shortage of tribological coatings to try for almost any need.

The primary problem that exists is in knowing what coating to select for any given application and what tests to run to evaluate potential coatings. A related problem is that coating developers often do not know where their coatings should be used or what coatings to develop to meet a particular need. There is a need for a strategy or methodology for selecting a coating for a given tribological application. In this paper such a strategy is proposed and the elements of that strategy are discussed.

II. Strategy The proposed strategy is outlined in Table III. The process begins with outlining the coating requirements. Next the coating function must be defined. A list of tribological functions is given in Table IV. From the coating requirements and function, the properties required of the coating can be defined. Often these are only qualitative; however, in some cases precise values can be assigned. The types of properties are listed in Table V. The coating composition may then be selected. Once the composition is selected the coating application technique can be chosen. The application technique is also determined by the component parameters (size, shape, material, subsequent machining required, and cost) and by the control required in the coating (thickness, composition, purity). Tests to evaluate the coating must then be determined. It can be seen that three types of tests can be run: Tests to evaluate the tribological function, property tests, and component performance tests. These will be discussed in a later section. The key to this strategy is understanding the tribological function. Unlike corrosion resistant coatings which have only one function (prevention of corrosion), tribological coatings are applied for a variety of functions.

III. Tribological Functions Ten tribological functions are listed in Table IV. Coatings may be applied to reduce adhesion and adhesive wear. Low adhesion is attributed to nonwelding, insoluble, or nonwetting materials combinations. Metal/oxide combinations such as silver/aluminum oxide show low adhesion. Polymer/metal combinations also have low adhesion. A difference in bond type seems to be required. To reduce cutting wear, high hardness is required. High friction also reduces cutting wear. Low deformation wear requires higher tensile strength, greater tensile elongation and lower elastic modulus.

Lubricant retention is maximized by porous surfaces such as are found with phosphating treatments. Certain polymers such as nylon appear to hold lubricants better than metals. Improved load capacity is not well understood yet but soft coatings with good heat dissipation and low friction are good performers. There is a critical failure temperature for lubricants and any means to reduce local surface temperature will increase load capacity. Soft films which distribute the load, especially those which polish under sliding conditions, show good tribological properties. "Enhanced lubricant effectiveness" refers to use of a coating which is more easily lubricated than the base metal. Coating titanium, which is difficult to lubricate, with copper or bronze is an example of this function. Rebuilding surfaces means adding a coating for dimensional purposes. Coatings are also used to either increase or decrease friction. Molybdenum disulfide and teflon coatings reduce friction while certain abrasive-containing coatings increase friction. It is extremely important to recognize these different functions when selecting or developing coatings.

It is not always necessary to select a coating based upon properties since certain compositions are known to fulfill certain functions better than others. Table VI shows examples of coatings that apply to certain functions. Some general conclusions can be drawn from this table. Most of the coatings which have been developed emphasize hardness and much of the research being done emphasizes increased hardness for wear resistance. Yet this property is only required in a few of the listed functions. Furthermore, some of the other functions are not adequately covered by compositions involving high hardness. For example, new compositions are needed to improve fracture toughness, to enhance fluid film lubrication, and to provide lubricant additives at the contact surface.

IV. Laboratory Tests for Coatings An essential element in the strategy presented here involves the use of suitable laboratory tests to evaluate coating performance prior to actual service applications. Because of the usual complexity of wear modes in service, it will always be necessary to evaluate several coating choices. As shown in Table IV, many different tribological functions exist and several of them usually combine in any service requirement. Laboratory tests for coatings can be considered in three categories (Table VII): wear tests, properties tests, and simulated operation tests. The first two are considered bench tests and measurements; they involve relatively simple and possibly standard configurations. The third type of test is established to simulate some of the more complex conditions of the final application, and frequently involves operation of actual components under controlled but relevant conditions. This sequence of testing is a necessary part of the strategy discussed earlier (Table III) and should be used in any coating development or coating selection process.

Coating properties measurements include such topics as adhesion strength (to the substrate), thermal expansion, hardness, and pore and crack content. Several of these have been discussed in earlier papers in this conference (see Hintermann[1], for example) and will not be described here. Rather we shall elaborate on possible bench wear tests for evaluation of coatings. Table VIII lists several standard wear tests that are available. They have the advantage over a great number of other published wear tests in that a substantial data base exists in these cases for comparison. Under either dry or lubricated metal-to-metal sliding conditions, the block-on-ring test (ASTM G77-83) and the crossed cylinder tests (preliminary ASTM standard) can be used to evaluate wear and friction proprties of coatings against a number of common alloys (e.g., 52100 bearing grade steel) or in self-mated tests. Data obtained from interlaboratory comparison tests can be obtained from ASTM* and

*Write to ASTM, 1916 Race Street Philadelpia, PA 19103, requesting the research report on each test of interest.

139

used to verify the correct operation of equipment and procedures, thus providing confidence in extending the tests to new, unknown materials. Table IX lists the standard conditions and some examples of wear values for the crossed-cylinder test. Two standard abrasion tests are also listed, one dry test involving low stress three-body abrasion (G65-81), and the other concerned with water slurry abrasion. The dry abrasion test has been studied in an interlaboratory comparison and found to be very reproducible (Table X). In some applications involving abrasive wear, the conditions may be so different that these particular standard tests may be inappropriate. In that case, a more specialized laboratory bench test may need to be developed. However, in many cases involving abrasive wear in industry, it has been noted that the dry sand/rubber wheel test provides a ranking of materials consistant with their relative performance in use. Hence it can be argued that use of such a bench test in a materials development program for abrasive resistant materials has broad merit.

It is clearly desirable to develop more standard bench wear tests, so that the test selected is as appropriate as possible for the coating problem at hand. In fact, none of these tests are specifically designed for coatings work. A rational approach is as important here as in the process development area. In view of the expense of full scale, component testing, the cost-effective route for coating development and selection requires the availability of well established and relevant standard wear tests (and properties tests).

The third essential category of laboratory tests for coatings, simulated operation tests, represents an intermediate step between full scale service testing and bench testing of small specimens (as contrasted to components). Examples of tests in this category include engine dynamometer testing of lubricants and materials used in internal combustion engines. This stage in the strategy is likely to uncover oversights in previous testing approaches or in the bench tests themselves. For example, consider an application where the coating function involves temperature cycles accumulating to a substantial number over the intended lifetime. Bench testing under certain temperature cycling conditions may overlook some key aspects, such as time at the highest temperature (where oxidation or diffusion processes are rapid), or the maximum rate of change of temperature during the expected service. As a result, coating/substrate combinations that perform well in the bench testing stage may not survive more representative simulated service condition testing. Guidelines for developing an adequate simulated test are difficult to formulate in general terms. It may be necessary to conduct parallel tests, for example, wear tests and thermal cycling tests, and then attempt to infer combined performance from the separate results. Careful analysis of all the conditions expected in service is essential if simulated testing is to be meaningful.

V. New Classes of Coatings The development of new coatings has proceeded at a rapid rate for many years, as evidenced by the literature in this area and also by the frequency of technical conferences on coatings (see for example, the proceedings of the annual American Vacuum Society meeting on the subject; ref. 2). A recent literature search by the authors revealed 1100 references concerned with tribological metallic coatings and surface treatments, reported since 1965. A few trends have emerged in recent years that involve new technologies. Two of these areas are shown in Table XI; directed energy sources and ion implantation. Both lasers and electron beam systems of sufficient power are being used to surface anneal and also to surface melt various alloys. Pulsed systems and continuous duty systems are used in either case, depending on such details as specimen size, affected depth desired, metal types, etc. If the processing involves surface melting, rapid solidification results afterward and this can lead to new phases and structures, even to amorphous layers on normally crystalline substrates. The use of directed energy sources involves motion of the beam across the surface being treated

140

as shown in Fig. 1. A technique developed at the Naval Research Laboratory[3] further modifies the system to include a nozzle that directs a stream of hard particles at the molten metal pool. This produces a metal matrix composite surface region that has shown very good abrasive wear behavior in a study[4] of aluminum and titanium alloys (Fig. 2). With this approach, surface alloying is also possible. For example, elements such as chromium and vanadium could be alloyed into the surface region of a conventional tool steel to improve its high speed cutting performance. Electron beam sources can be used in much the same way to process the surface of metals. Since that approach involves a vacuum chamber, contamination and oxidation effects are much reduced. In a study of the effects of electron beam surface melting on tool steel wear performance at the National Bureau of Standards[5], improvement in wear rate (dry sliding, O-2 vs 52100 steel) by a factor of nearly 4x was achieved relative to normal quenching and tempering treatments. As shown in Table XII, that improvement was not due to changes in steel hardness. Microstructure studies revealed that rapid solidification of the electron beam melted surface produced a fine sized, tempered, twinned martensite structure that was more resistant to wear.

Ion implantation[6,7] involves the injection of high energy ions into the near surface region of a metal, producing altered compositions and new phases in some cases. Wear performance improvements have been observed, and are attributed in the case of steels implanted with titanium to increased hardness at depths up to 100 nm. Tribological performance can be affected in other ways than by hardness changes. A recent study[8] of ion mixing of tin into Ti-6A1-4V alloys showed an improved wear performance due to reduced adhesive effects in the sliding contact zone, attributed to the formation of a tin-rich film. Through the ion mixing technique, it may be possible to improve the tribological properties of many alloys by creating an internal source of atoms (implanted) that will become available for film formation within the contact zone under mild wear condtions. Friction properties may also be tailored by such selective ion implantation.

VI. Conclusions This approach to the development and selection of tribological coatings involves a specific strategy. The coating function must be accurately defined in all respects, and laboratory tests at several levels must be completed. These tests determine appropriate wear performance and measure coating properties to insure that the function can be achieved. Standard wear tests are employed as much as possible in ranking candidate materials and processing techniques. Simulated operation tests are then necessary to further evaluate the coating performance. It is argued that such a systematic approach to the development of and the testing of tribological coatings is now feasible. In this way, classes of coatings and surface treatments can be tailored to particular types of applications in a cost-effective way.

References

1. H. Hintermann, Proc. Mechanical Failure Prevention Group, 37th Meeting (Cambridge Press, 1983).
2. Metallurgical Coatings, 1982 (Elsevier Sequoia, Lausanne, Netherlands).
3. J. D. Ayers, R. J. Schaefer, and W. P. Robey, J. of Metals, 33, 1981, pp. 19-23.
4. J. D. Ayers, L. K. Ives, F. Matanzo, and A. W. Ruff, Wear of Materials, 1983 (ASME, NY), pp. 265-271.
5. A. W. Ruff, J. S. Harris, Wear of Materials, 1981 (ASME, NY), pp. 326-334, and unpublished work.
6. J. K. Hirvonen, J. Vac. Sci. Technology, 15, 1978, pp. 1662-1668.
7. G. Dearnaley, in Ion Implantation Metallurgy, ed. C. M. Preece and J. Hirvonen (Met. Soc. AIME, PA, 1980), pp. 1-20.
8. G. Dearnaley, Metallurgical Coatings, 1983 (in press).

Table I   Tribological Compositions In Frequent Use

Case Hardening	Oxides
Nitriding	Phosphate
Boronizing	Steel/Co/Ni Hard Face
Sulfiding	Al_2O_3
Chromizing	ZrO_2
Chromium Plate	TiC, Other Carbides
Electroless Nickel	Teflon
Tin, Lead, Indium	Nylon
Copper, Bronze	Polyurethane
Lead-Tin-Copper	MoS_2/Graphite/Teflon (Bonded)
Silver Plating	Anodizing with Additives

Table II   Coating Processes

Diffusion	**Welding**
Gas	Oxy-Acetylene
Liquid	Tungsten Inert Gas
Pack	Shielded Metal Arc
	Open Arc
Thermal	Metal Inert Gas
Induction Hardening	Submerged Arc
Flame Hardening	Electro Slag
Laser Hardening	Paste
Electron Beam Hardening	Plasma Arc
Spark Hardening	
	Spraying
Plating	Flame
Electroplating	Electric Arc
Electroless Plating	Plasma
Brush Plating	Detonation
Laser Electroplating	
Pulsed Electroplating	**Physical Vapor Deposition**
	Evaporation
Painting	Sputtering
Air Dry	Reactive Sputtering
Heat Cure	Ion Plating
	Ion Implanation
Transfer Films	
Rubbing	**Chemical Vapor Deposition**
Tapes	Flowing Gas
Putties	Laser CVD
Fusion	**Chemical Reaction**
Glazes	Oxidation
Powder	Anodizing
	Acid Bath

Table III Strategy for Coating Selection and Development

Table IV Tribological Functions

Reduce Adhesion & Adhesive Wear	Increase Load Capacity
Reduce Cutting	Promote Film Lubrication
Reduce Deformation Wear	Enhance Lubricant Effectiveness
Reduce Corrosion Wear	Rebuild/Replace Surface
Retain Lubricant	Modify Function

Table V Properties Needed and Considerations

Wear Resistance
- Conditions, e.g., load, velocity
- Geometry, e.g., rolling,......
- Lubrication

Mechanical Requirements
- Strength Level
- Fatigue Resistance
- Impact Resistance

Chemical Resistance
- Type of Environment
- Exposure Limits, e.g., temperature

Table VI Functions of Tribological Coating Classes

COATING CLASS \ FUNCTION	Adhesive Wear	Cutting Wear	Deformation Wear	Corrosion Wear		Lubricant Retention	Load Capacity	Fluid Film	Improve Lubrication		Rebuild Surfaces		Modify Friction
Case Hardening		0	0										
Nitriding		0	0										
Boronizing		0											
Sulfiding				0			0						
Chromizing				0									
Chromium Plate		0		0		0			0		0		
Electroless Nickel		0		0			0				0		
Noble Metals	0						0	0					
Oxides	0	0				0							
Phosphate						0							
Steel/Cobalt/Nickel Hard Face		0	0								0		
Al$_2$O$_3$		0											
TiC		0											
Mixed Carbides		0											
Teflon	0												0
Nylon						0							
Polyurethane													
Polypropolyene Sulfide													
MoS$_2$/Graphite/Teflon Bonded Coatings	0												0
Tin, Lead, Indium	0							0	0				
Copper, Bronze	0								0				

Table VII Laboratory Tests for Coatings

Wear Tests (e.g., abrasion
Properties Tests (e.g., toughness)
Simulated Operation (e.g., component tests)

Table VIII. Standard Wear Test Systems Suitable for Coatings

Dry Sand/Rubber Wheel Abrasion	Liquid Impingement Erosion
(ASTM G65-81)	(ASTM G73-82)
Block-on-Ring Sliding Wear	Vibratory Cavitation Erosion
(ASTM G77-83)	(ASTM G32-77)
Slurry Abrasivity	Solid Particle Impingement Erosion
(ASTM G75-82)	(ASTM G76-83)
Crossed-Cylinder Sliding Wear	
(ASTM Preliminary)	

Table IX. Proposed Standard Crossed Cylinder Wear Test

Load:	71.2 N (16 lbf)
Pressure:	1700 MPa (240 ksi)
Speed:	6.6 cm/s (100 rpm), or
	26.6 cm/s (400 rpm)
Distance:	400 m, or 1600 m, or 3200 m
Wear Values:	Self Mated Tests
	M4 steel ----------- 1.0 mm^3
	(71 N, 26.6 cm/s)
	3200 m
	304 steel ---------- 9 mm^3
	(71 N, 6.6 cm/s)
	400 m
	1020 steel --------- 210 mm^3
	(71 N, 6.6 cm/s)
	400 m

145

Table X. ASTM G2 Interlaboratory Test (G65-81)

| Laboratory | Wear Volume (mm^3) | |
	Average	Coef. of Var., %
A	34.8	6.7
B	35.2	2.1
C	36.9	5.9
D	38.7	4.3
E	36.7	2.8
Avg.	32.9	3.3

Conditions: D2 steel, AFS sand, 130 N load, 6000 rev.

Table XI. New Classes of Coatings

Directed Energy Sources	Ion Implanation
- Laser	- Single Element
- CO_2, CW or pulsed Systems	- Multiple Elements
- Surface Annealing	- Ion Mixing
- Surface Melting, RSP	
- Particle Injection	
- Alloy Modification	
- Electron Beam	
- Vacuum Enclosure	
- High Coupling Efficiency	
- Pulsed or CW	

Table XII. Wear Performance of Electron Beam Surface Melted Tool Steel

Alloy	Hardness (KHN_{1000})	Hardness Ratio	Wear Rate+ (mm^3/m)	Wear Ratio
0-2 (annealed)	230	1	1.6×10^{-4}	1
0-2 (quenched)	730	0.32	1.1×10^{-4}	0.69
0-2 (e-beam)	800	0.29	0.30×10^{-4}	0.19

+wear volume/total sliding distance vs 52100 steel, dry, argon, 10 N

146

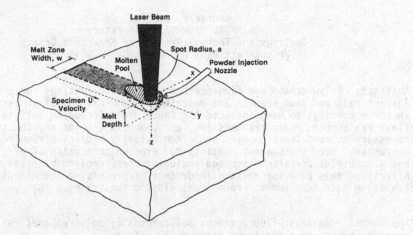

Figure 1. Schematic representation of system
used to inject particles into
molten surface region of alloys.

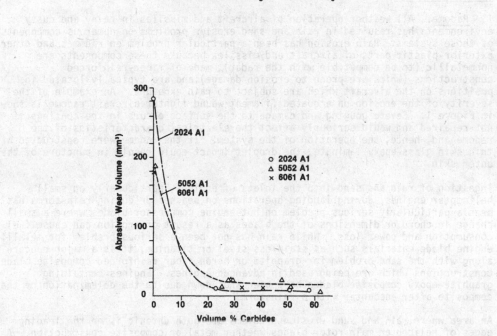

Figure 2. Reduction in abrasive wear rate of
aluminum alloys due to TiC particle
additions.

147

COATINGS FOR EROSION RESISTANCE

George F. Schmitt, Jr.
AFWAL Materials Laboratory
Coatings and Thermal Protective Materials Branch
Wright-Patterson Air Force Base, Ohio 45433

Abstract: Polyurethane and fluorocarbon elastomeric coatings for protection against rain and sand erosion are described. Combinations of properties required in these coatings to meet advanced Air Force needs including antistatic, thermal flash resistance, radar transmission, and camouflage color and the tradeoffs necessary to meet these complex and often mutually exclusive requirements are discussed. Hard transparent coatings for erosion protection of aircraft canopies and windshields are also discussed including their erosion behavior, ultraviolet effects on this behavior and the importance of processing and cleanliness in achieving good adhesion to transparent plastic substrates.

Key Words: Coatings; fluorocarbon; polycarbonate; polyurethane; rain erosion; sand erosion; transparent coatings; ultraviolet effects.

Erosion Problems:

I. Radome. All weather operation of aircraft and missiles in rainy and dusty environments has resulted in rain and sand erosion problems on numerous components of those systems. Rain erosion has been a particular problem on radomes and other exterior plastic parts of aircraft and missiles because these components are nonmetallic (to be compatible with the radar), made of fiber-reinforced constructions (which are prone to erosion damage) and are typically located in positions on the aircraft which are subject to rain exposure. An example of the severity of the erosion on a coated filament wound fighter aircraft radome is shown in Figure 1. Severe gouging and damage to the surface occurs in the coatings if not repaired and would seriously affect the electrical characteristics of the radome and, hence, the operation of the system. If the radome were constructed of thin skin glass-epoxy laminate, the droplet impact could result in puncture of the outer skin.

Ingestion of rain and sand into the inlets of engines, particularly on small helicopter engines, during landing operations on beaches or during rainstorms has been a particularly serious problem on hot engine compressor blades where a small change in chord or dimensions of the blades as a result of erosion can cause fuel consumption and power loss. While rain has not been a serious problem for metallic engine blade materials such as stainless steel or titanium, it is a major concern along with the sand problem for graphite or boron fiber reinforced composite blade constructions which are being used in advanced engines. Engines containing graphite-epoxy composite blades have been shut down due to the delamination of the composite after encounter with a rainstorm.

An area where rain and sand erosion problems are both chronic is on the leading edges of helicopter main rotor blades whether metal or composite construction. A variety of protective materials schemes have been attempted with some success, but usually with a severe weight penalty.(1)

The potential erosion problems with thin skin construction are particularly important for electronic countermeasure antenna covers where broad frequency band requirements dictate minimum skin thickness in order to achieve sufficient and efficient jamming/obscuration operations. Honeycomb constructions are most often employed and outer skin thicknesses of 0.030 inch minimum are recommended to avoid puncture due to droplet impact. Care must also be taken to insure that the inner skin of a honeycomb is rigid enough to prevent spallation when the impact loads from the droplet striking the outer surface are transmitted to the inner skin.

In actual service, an ECM antenna cover was constructed of glass-epoxy honeycomb with a single glass ply on the inside and when the droplets impacted on the polyurethane-coated outer surface, the inner ply spalled off and the part became unserviceable. In other service experience, crushing of the honeycomb and holes through the outer skin have been experienced.

All of the above problems are primarily subsonic in nature. As the velocity increases to a supersonic speeds, the erosion problem is particularly compounded. Pilots are presently instructed to avoid rain at supersonic speeds. However, for tactical aircraft and high speed missiles, it is desirable and necessary to be able to penetrate rainstorms at supersonic velocities, especially for low level dash missions. Unprotected parts of aircraft and missiles such as radomes exposed to rain supersonically will be destroyed in a few seconds. To overcome this, ceramic caps or all-ceramic radomes could be used with or in lieu of plastic radomes. However, these are limited because of inherent brittleness, thermal mismatch problems, attachment difficulties, and structural weaknesses of the ceramics.

Transparencies: An increasing number of high performance aircraft are being fitted with transparencies utilizing polycarbonate material as the structural ply. This usage is dictated by the need to provide a transparency which can survive high speed impact energies. The impact resistance of polycarbonate material is influenced by such parameters as thickness, temperature, configuration, surface finish, aging and environmental exposure. Coated monolithic polycarbonate transparencies, as well as laminated acrylic/polycarbonate plies, are considered as current and potential future applications. In either case, outer and inner surface protection of the polycarbonate must be provided by protective coatings or acrylic plies.

Investigation of the rain erosion behavior of transparent plastics, specifically polysulfone and uncoated polycarbonate, has been conducted by the Air Force Wright Aeronautical Laboratories (AFWAL) Materials Laboratory and reported. [2] Although numerous proprietary coatings for polycarbonate have been evaluated in the AFWAL Materials Laboratory rotating arm apparatus over the past seven years, the results in this paper represent the first systematic study of the influence of environmental exposures on the erosive behavior of those transparent coatings.

II. Aircraft Radome Coatings. Protection of aircraft radomes in the Air Force utilized MIL-C-7439B black and MIL-C-27315 white neoprene coatings, which were developed by AFML in the 1950's up until about 1969, but they suffered from poor weatherability, difficult application procedures and provided inadequate subsonic erosion resistance for recent and more advanced aircraft. The neoprenes were also limited in service temperature to 90°C or below.

Initial inhouse research at AFML in 1965 had indicated that polyurethanes as a class appeared to show promise for improved rain erosion resistance. As a result, research efforts were sponsored to develop new and improved polyurethane coatings

149

specifically tailored for subsonic rain erosion protection. As a result of this research, protection of aircraft radomes in the Air Force currently utilizes MIL-C-83231A and MIL-C-83445A two component or moisture cure polyurethane rain erosion resistant coatings. The outstanding performance of the polyurethane in actual service has eliminated erosion problems on many aircraft and service times in excess of three years for these polyurethane coatings are routinely achieved. However, the polyurethanes are limited in service temperatures to 150°C for long times and 175°C for shorter periods (up to 24 hours). At temperatures above 200°C, the polyurethane erosion coatings lose their elastomeric character and their erosion resistance very rapidly.

Fluorocarbon coatings based on vinylidene fluoride-hexefluoropropylene copolymers retain erosion resistance after long term exposure to 260°C. The fluoroelastomer coating withstands rainfall conditions of 25 mm/hr at 223 m/s for an average 100 minutes. The coatings demonstrate one way power transmission of 94 percent at 9.275 GHz and surface resistivity of 0.5 - 15 megohms per square. They can be applied to full 0.3 mm thickness in 3 hours using conventional spray techniques, cure at room temperature, and maintain erosion resistance, transmission and antistatic properties after outdoor exposure of three years or longer.

Recent developments in the incorporation of conductive fibers in the topcoat of the fluorocarbon coatings has now provided a technique for antistatic protection without requring the use of carbon black (previously the state-of-the-art) with a resulting black color for the radome. Camouflage of the aircraft radome is desirable and the fibers enable the base erosion coatign to be tailored for whatever color (gray, brown, green) is needed for matching purposes. These fibers are used at a low level of loading so that radar transmission and erosion resistance are maintained. Unfortunately, the fibers cannot be employed in the polyurethane coatings due to inability of the elastomeric polyurethane to wet out and adhere to the fibers; as a result, erosion resistance is reduced to an unacceptable level.

With the use of the MIL-C-83231A and MIL-C-83445A polyurethane coatings for rain erosion protection of aircraft systems where the temperature does not exceed 150°C for long periods and the use of the new fluoroelastomer coating for higher temperature radome erosion protection, these two coatings provide a significant capability for aircraft radome subsonic rain erosion resistance. Table 1 summarizes the properties of these various coatings. References 3 and 4 describe solid particle erosion mechanisms and liquid impact erosion mechanisms in these coatings.

Environmental Laboratory Testing. One of the keys to developing and optimizing the coatings which exhibited the greatest erosion resistance was the extensive use of the rotating arm apparatus to characterize the dynamic response of these coatings to the rain environment. The rotating arm was located at the Air Force Wright Aeronautical Laboratories/Materials Laboratory. A schematic is shown in Figure 2. The apparatus consists of an eight-foot diameter double arm blade. It is designed to produce high tip velocities with negative lift and a low drag coefficient. Mated test specimens are mounted at each leading edge tip section of the double rotating arm. The test specimens can be subjected to variable speeds of 0 to 900 mph. The double arm is mounted horizontally on a vertical drive shaft. Simulated rainfall is produced by four curved manifold quadrants. Each manifold quadrant has 24 equally-spaced capillaries. Raindrop size and drop rate are controlled by the capillary orifice diameter and the head pressure of the water supply. The manifold quadrants are mounted above the tips of the double rotating arm. Raindrops from the simulation apparatus impact the test specimens throughout their entire annular path. Rain droplets are 2.0 mm diameter and generated at the rate of one inch/hour of simulated rain fall.

The specimens are exposed in this environment until failure of the coating by penetration to the substrate, which is determined by observation of the specimens while running through use of a stroboscopic light and closed circuit TV camera. The performance of coatings in the rotating arm apparatus has been correlated to actual flight test results both as to ranking of various materials and the actual modes of failure of the materials themselves.

Outstanding erosion resistance is not the only requirement for a coating to be used in actual service. Other requirements include: (a) good dielectric transmission properties for radar transmission; (b) antistatic properties for reduction of precipitation static on large radomes; (c) good weatherability including retention of rain erosion resistance, transmission and surface resistivity after prolonged outdoor exposure; (d) application techniques and cure properties which are compatible with repair of radomes at field activities and (e) thermal flash resistance for protection against nuclear blast radiation. Repair of the coatings themselves as well as strippability, if the coating must be removed, are also required.

The screening data on new coatings on appropriate substrates and the information on performance of various composite/honeycomb/lay-up radome constructions from the rotating arm apparatus is also obtained on artificially weathered (i.e., Weatherometer) and Florida-weathered specimens to determine the weatherability and retention of properties of these coatings in service. These tests help to provide assessment of whether a conductive p-static protection outer coating will become non-conductive as it weathers on the radome, thus no longer providing antistatic protection. Changes in transmission through a weathered coating also are determined but in general with the polyurethane and fluorocarbon coatings have been found to be a minor effect. Similarly, key properties changes after weathering are determined with rotating arm tests, thermal flash tests, etc.

Despite the confidence in the relative ranking of materials erosion performance on the rotating arm due to the correlation tests which were conducted on an F-100 aircraft with materials of varying erosion resistance which verified the rankings, flight tests of coatings on radomes are essential. The ability of the rotating arm apparatus to provide an accurate assessment of a coating's capability is further reinforced by the outstanding performance of coatings such as the MIL-C-83231 polyurethane which was predicted to be at least 5 times as resistant as the old MIL-C-7439B neoprene based on as applied and weathered specimens and in actual service has been even better than that.

III. Transparent Coatings. The purpose of the tests reported in this section was to determine the relative rain erosion resistance and damage mechanisms involved with different types of coated monolithic polycarbonate transparency materials. The materials were to be evaluated before and after simulated environmental exposure. Solid particle erosion of transparent coatings on polycarbonate has been previously reported. (5)

Target Materials. All materials for evaluation were supplied by the U.S. Air Force. The materials were furnished in flat sheet form and processed to be representative of material used in aircraft transparencies. The three vendors supplying coated monolithic polycarbonate material are identified as Vendor P,

Vendor A, and Vendor B. Three different coatings were utilized: Vendor P material was supplied with the Texstar C-254-1C coating, Vendor A had its own specific coating, and, similarly, Vendor B had its own specific coating.

Simulated UV Exposures. Accelerated UV Laboratory Conditioning
The ultraviolet radiation environmental conditioning of the test specimens was implemented in two stages: the as-received baseline materials consisting of no UV exposure, and the baseline materials subjected to one and three years of accelerated laboratory UV exposure. All ultraviolet conditioning was performed using a "Sunlighter IV" accelerated sunlight tester, manufactured by the Test-Lab Apparatus Co., Amherst, New Hampshire. Basically, this apparatus consists of four sunlamp bulbs mounted over a rotating turntable. The tester acceleration ratio over natural sunlight is based on a cabinet temperature of 131°F - 140°F. The energy level in the range where nearly all UV degradation occurs, supplied by the General Electric RS-4 sunlamp bulbs in the tester, varies from a wavelength of 290 millimicrons (nanometers) at an intensity of 1300 watts/sq. meter to 360 millimicrons at 30,000 watts/sq. meter. The wavelength of maximum sensitivity for polycarbonates being 295 millimicrons. Specimens were mounted on a screen to avoid contact with the nonreflective turntable.

One sunlamp bulb is mounted directly over the center portion of the turntable, and three additional bulbs are mounted over the outboard portion of the turntable. Consequently, two areas with different exposure accelerations are produced on the turntable, an inner circle of approximately six inches diameter, and the remaining outer ring of 17.5 inches diameter. For the inner circle, the acceleration ratio is approximately eight hours exposure equivalent to one year natural sunlight according to the manufacturers. The inner circle was used for all UV exposure of the rain erosion test specimens. For the purposes of this paper, the data will indicate no accelerated UV weathering, one year UV accelerated weathering, and three year UV accelerated weathering, respectively.

Accelerated Outdoor Sunshine Conditioning. Accelerated outdoor weathering of simulated three-year exposure was accomplished by utilizing the Equatorial Mount with Mirrors for Acceleration (EMMA) machine at the Desert Sunshine Exposure Test (DSET) Laboratory near Phoenix, Arizona. It is estimated that 40 days of exposure on the EMMA machine is approximately equivalent to one year of exposure to 45-degree south natural weathering. The specimens received about eight times as much radiation as those exposed on a follow-the-sun rack during equal periods of time. Each simulated year was based on an exposure rate of 164,250 langleys.

Rain Erosion. For the purposes of this study, matched specimens of the coated monolithic polycarbonate materials were inserted into the leading edge tip sections of the double rotating arm at a 30° angle of incidence to the rain droplet impact. All rainfield exposure testing was conducted at 500 miles per hour in a calibrated one inch per hour simulated rainfall. Duration of the tests was established at 1, 2, 5, 10 and 15 minutes intervals.

Visual Observations. All specimens were examined after rainfield exposure with an illuminated magnifier and the surface condition was recorded. Comments included scratches, pitting and percentage of coating removal.

Surface Characterization. The surfaces of the exposed specimens were examined directly by scanning electron microscopy (SEM) together with an x-ray unit. Specimens were vapor shadowed with a heavy metal or carbon to provide contrast.

Hazemeter Measurements. Haze and transmission measurements were determined using the Standard Test Method for Haze and Luminous Transmittance of Transparent Plastics ASTM Method D 1003. These methods were found to be too insensitive to changes in the coatings and were abandoned.

Results. The primary effects of the rainfield exposure was examined. Namely, the rain erosion kinetics (percent coating removal as a function of accelerated UV exposure and duration of rainfield exposure) and the surface morphology of the coated transparencies after rainfield exposure.

Effect of Accelerated UV Exposure. The Vendor P specimens (one year and three year accelerated UV exposure) exhibited a decrease in the erosion incubation period and an increase in percent coating removal as a function of increasing accelerated UV exposure, with a decrease in rainfield exposure time as shown in Figures 4 and 5. The Vendor A specimens (one year and three year accelerated UV conditioning) demonstrated no decrease in the erosion incubation period, but exhibited a marked increase in percent coating removal as a function of UV weathering with increasing rainfield exposure duration. Vendor B specimens exhibited severe pitting of the coating surface at the 15 minute rainfield exposure interval, but there was no change as a function of the amount of accelerated UV weathering.

Scanning Electron Microscopy Examinations. Selected test specimens which were exposed for varying time intervals to a 1 inch/hour rainfield at 500 mph were examined by scanning electron microscopy (SEM) for characterization of the mode of erosion damage and erosion processes. A typical eroded surface of the coated monolithic polycarbonate transparencies is shown in Figure 6. The formation of pits and associated crack propagation in the coating film is clearly observed.

The Vendor A coating as compared to Vendor P appears to have higher ductility. This may be deduced by comparing Figure 7 (where coating foldback was possible due to high coating ductility) with Figure 8 (where coating brittleness is demonstrated by the formation of sharp boundaries and cracks in the zones of removed coating). Vendor A coatings could absorb more rain droplet impact energy and remain adhered to the polycarbonate substrate, resulting in a longer incubation period as well as a decreased coating removal rate.

Coating removal in the Vendor B material was nearly zero percent, being independent of rainfield exposure time as well as UV radiation exposure effects. The lack of substantial coating removal may be associated with the existence of high adhesion forces between the coating and substrate interface. This may be deduced from Figure 9, where the coating itself was eroded prior to delamination from the substrate surface.

The difference in adhesion of the various coatings is illustrated in Figure 10 for Vendor P, Figure 11 for Vendor A and Figure 12 for Vendor B coatings, respectively.

Effects of Accelerated UV Exposure. Accelerated UV exposure effects on the subsequent rain erosion behavior of the Vendor P, A, and B materials were evaluated as a function of a kinetic behavior of coating removal and surface characterization as obtained by SEM observations. Exposure of the Vendor P materials to accelerated UV weathering resulted in reducing the erosion incubation period from 2 minutes to 0.5 minutes with a significant increase in the rate of coating removal. The erosion incubation period of the Vendor A materials was not affected by accelerated UV weathering exposure; however, coating removal rate did substantially increase with UV exposure as shown by comparing Figures 3 and 4. Further, more exposure to accelerated UV weathering apparently caused the introduction of damage into the polycarbonate substrates of Vendor P and Vendor A, in the presence of moisture. Consequently, the resultant reduction in adhesion of the coatings to the polycarbonate substrates at their interface led to the decrease in incubation time as well as an increase in the rate of coating removal in the Vendor P and Vendor A materials.

Since the polycarbonate substrates were apparently affected by accelerated UV weathering, it is suggested that the Vendor P and Vendor A coatings were transparent in the UV wavelength region. Whereas in the Vendor B material, essentially no coating removal was visually observed as well as no apparent damage to the polycarbonate substrate as shown in Figure 9. These observations suggest that Vendor P and Vendor A coatings did not apparently contain the necessary type and amount of UV absorbers. The Vendor B coating may have contained sufficient amounts of UV absorbers which apparently prevented damage to the polycarbonate substrate and consequently prevented delamination and coating removal. Accelerated UV weathering did not substantially affect the coating properties in so far as their rain erosion properties is concerned. In all three cases, the coatings were subject to rain erosion damage prior to accelerated UV weathering. Similar rain erosion damage occurred after exposure to accelerated UV weathering. These observations support the hypothesis that accelerated UV weathering did not directly affect the coatings behavior under rainfield exposure conditions, but rather had an indirect effect through the reduction of adhesion to the polycarbonate substrate as discussed above.

Effects of Desert Sunshine Exposure. Exposure of the Vendor P material to Desert Sunshine EMMA test, previously described, for the equivalent of three years and thereafter to rainfield exposure conditions resulted in coating removal as shown in Figure 13. Damage was also evident in the polycarbonate substrate as shown in Figure 14. The mode of coating removal was very similar to the Vendor P specimens exposed to accelerated UV weathering in the laboratory. Furthermore, the mode of damage introduced in the polycarbonate substrate was associated with the formation of craters together with microcracks, as was the case for the laboratory UV irradiated Vendor P materials. Based on these findings, it is possible to deduce a good correlation between the rain erosion behavior of the Vendor P laboratory UV irradiated specimens and the Vendor P desert sunshine exposures.

Rain Erosion Processes. The rain erosion processes observed in the Vendor P, A and B specimens, UV irradiated and non-UV irradiated, can be characterized through two basic stages: (a) initiation process and (b) propagation process.

Rain Erosion Initiation - Surface Coating. The initiation process was associated with the incubation time. The erosion process at the initiation stage was characterized by localized erosion events which results in isolated craters or pits in the range of 10 to 30 m in size. These localized events consisted of material removal from the coating as well as localized coating separation from the polycarbonate substrate. With continued rainfield exposure, these local isolated events merged with each other.

Rain Erosion Initiation - Polycarbonate Substrate. Further exposure of the test specimens, beyond the incubation period, resulted in rain erosion damage in the polycarbonate substrate material. Substrate damage occurred in the UV irradiated test specimens of the Vendor P and A substrates but not in the Vendor B substrates. The initiation stage of rain erosion damage was characterized by the formation of localized craters or pits, up to 10 microns in diameter, with associated microcracks.

Rain Erosion Propagation - Surface Coating. The propagation stage of the rain erosion process, affecting the surface coatings, occurred through the joining of the local damage events. Continuation of the damage "growth" process of the local events resulted in coating removal from large areas a few hundred microns in size, and subsequently from the entire area of the test specimen exposed to rainfield conditions. Coating removal through the propagation stage correlated with the high rate of coating removal as shown in the S-type kinetic curves. Whenever the propagation stage of uniting local erosion events stopped, no general erosion damage, i.e., coating removal, occurred and the coating surface remained fundamentally undamaged as observed in the Vendor B materials. The observation that no advanced propagation stage occurred in the Vendor B material was also reflected in the kinetic curves, as the amount of coating material removal was essentially insiginificant.

Rain Erosion Propagation - Polycarbonate Substrate. The erosion propagation mode in the polycarbonate substrate could be characterized in several forms. One form suggest the combination of initially formed craters or pits, after growth to 30 microns in size, to form an elongated channel. Another form suggests the joining of minute isolated events (one micron in size) to form a continuous stream of parallel channels.

From the preceding discussion, it should be clear that the erosion effects on transparent coatings are quite subtle and quantitative differences were difficult to determine. Scanning electron microscopy was invaluable in determining differences in the coatings behavior. Conventional haze and transmittance tests were not sensitive enought to determine differences in these coatings.

The development of erosion resistant transparent exterior coatings for polycarbonate canopies and windshields remains a challenging task.

REFERENCES

1. G. F. Schmitt, "Liquid and Solid Particle Impact Erosion", Wear Control Handbook, Edited by M. B. Peterson and W.O. Winer, American Society of Mechanical Engineers, New York, 1980, pp 231-282.

2. G. F. Schmitt, Jr., "Rain Droplet Erosion Mechanisms in Transparent Plastic Materials", Proceedings of the Fourth International Conference on Rain Erosion and Associated Phenomena, Meersburg, Germany, 8-10 May 1974, A.A. Fyall and R.B. King, eds., Royal Aircraft Establishment, Farnborough, England.

3. J. Zahavi and G. F. Schmitt, "Solid Particle Erosion of Polymeric Coatings", Wear, Vol. 71 (1981), pp. 191-210.

4. C. J. Hurley, J. Zahavi and G. F. Schmitt, "Rain Erosion Mechanisms of Polyurethane and Fluoroelastomer Coated Composite Constructions", To be published in Proceedings of Sixth International Conference on Erosion by Liquid and Solid Impact, Cambridge, England, September 1983.

5. J. Zahavi and G. F. Schmitt, Jr., "Solid Particle Erosion of Coatings on Transparent Materials and Reinforced Composites", Selection and Use of Wear Tests for Coatings, ASTM STP 769, R. G. Bayer ed., American Society for Testing and Materials, 1982, pp 28-70.

TABLE I
COMPARISON OF AVAILABLE RAIN EROSION COATINGS

PROPERTY	MIL-C-7439B Neoprene	MIL-C-83231 Black Polyurethane	MIL-C-83445 White Polyurethane	AF-C-VBW-15-15*** White Fluorocarbon
*EROSION RESISTANCE MINUTES @ 500 MPH/1 in./Hr. RAIN	40	160	75	110
DIELECTRIC TRANSMISSION	EXCELLENT	EXCELLENT	EXCELLENT	EXCELLENT
SERVICE LIFE	<6 MOS.	>3 YRS.	>2 YRS.	>5 YRS.
TEMPERATURE LIMIT	200°F	300°F	300°F	500°F
EASE OF APPLICATION (ALL SPRAY APPLIED)	DIFFICULT	MODERATE	MODERATE	MODERATE
SPECIAL REQUIREMENTS	LOW HUMIDITY	40% R.H.	40% R.H.	NONE
STRIPPABILITY	EASY	EASY	EASY	FAIR
COST	LOW	MODERATE (COMPETITIVE ON COVERAGE BASIS)	MODERATE	HIGH
** THERMAL FLASH RESISTANCE CAL/CM² TOTAL FLUENCE	-	-	20 (ONE PULSE)	80 (AND MULTIPLE PULSES)

* AFWAL/ML ROTATING ARM APPARATUS
**DNA AERODYNAMIC QUARTZ LAMP WIND TUNNEL FACILITY
***NOTE: COLORED FLUOROCARBON COATINGS WOULD EXHIBIT SIMILAR PROPERTIES EXCEPT FOR THERMAL FLASH RESISTANCE.

FIGURE 1.   Eroded Radome

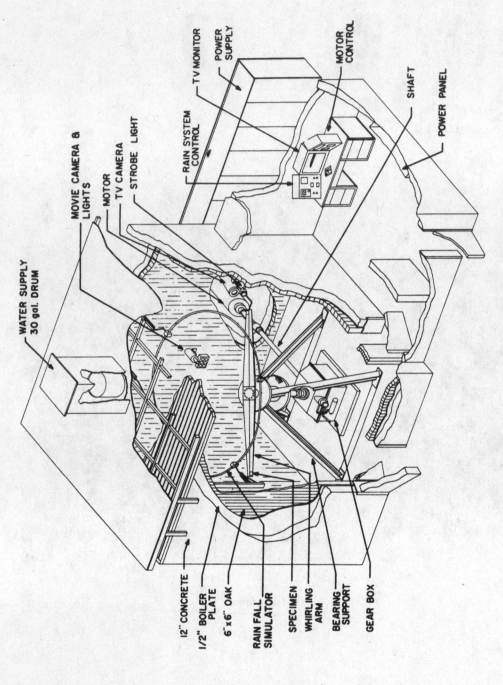

WATER SUPPLY 30 gal. DRUM

MOVIE CAMERA & LIGHTS

MOTOR

TV CAMERA

STROBE LIGHT

RAIN SYSTEM CONTROL

TV MONITOR

POWER SUPPLY

MOTOR CONTROL

SHAFT

POWER PANEL

12" CONCRETE

1/2" BOILER PLATE

6"x6" OAK

RAIN FALL SIMULATOR

SPECIMEN

WHIRLING ARM

BEARING SUPPORT

GEAR BOX

Figure 2. Mach 1.2 Rain Erosion Test Apparatus.

158

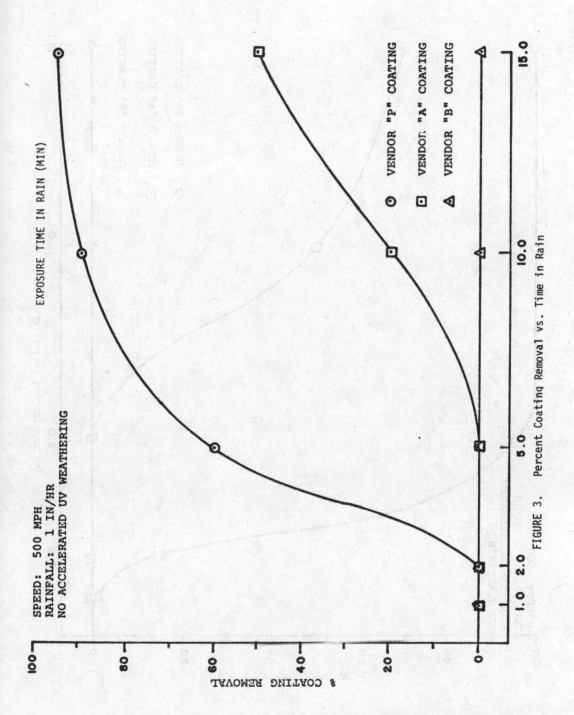

FIGURE 3. Percent Coating Removal vs. Time in Rain

159

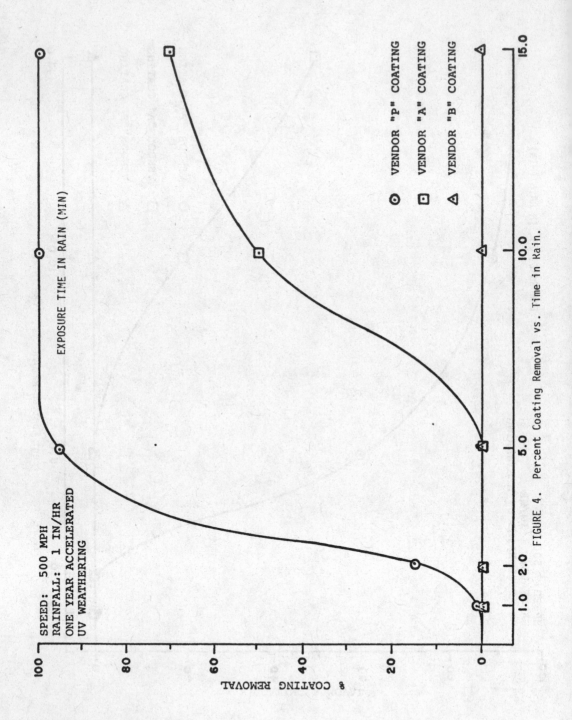

FIGURE 4. Percent Coating Removal vs. Time in Rain.

160

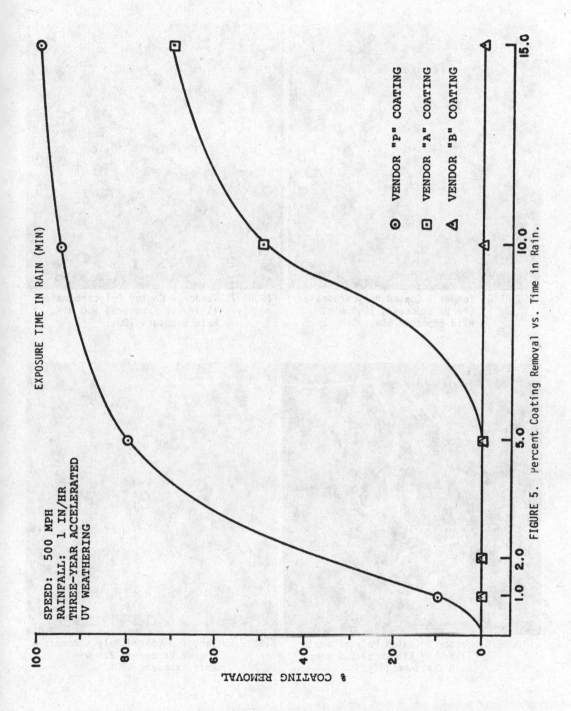

FIGURE 5. Percent Coating Removal vs. Time in Rain.

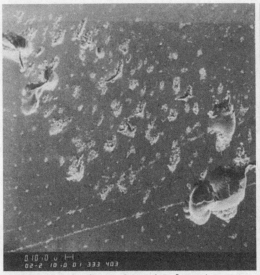 

FIGURE 6. Vendor B Coated Polycarbonate
(No UV Exposure) 15.0 min.
Rain Exposure 50x.

FIGURE 7. Vendor A Coated Polycarbonate
(1 Yr. UV Exposure) 5.0 min.
Rain Exposure 100x.

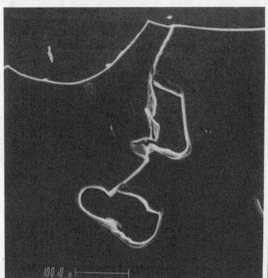 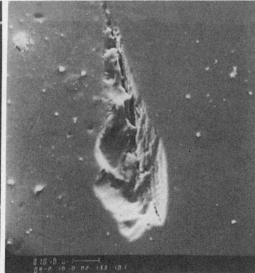

FIGURE 8. Vendor P Coated Polycarbonate
(1 Yr. UV Exposure) 2.0 min.
Fain Exposure 180x.

FIGURE 9. Vendor B Coated Polycarbonate
(No UV Exposure) 2.0 min.
Rain Exposure 800x.

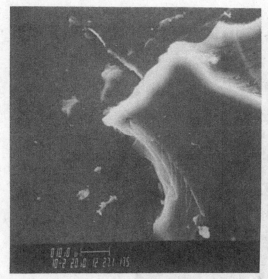

FIGURE 10. Vendor P Coating 15.0 min.
Rain Exposure. 1000x

FIGURE 11. Vendor A Coating 10.0 min.
Rain Exposure. 1000x

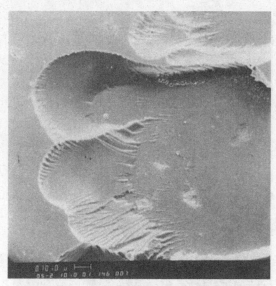

Figure 12. Vendor B Coating 10.0 min.
Rain Exposure. 500x

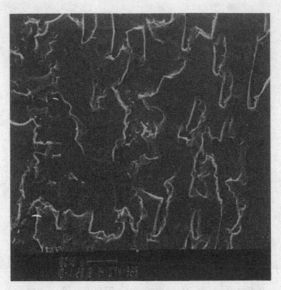

FIGURE 13.   Vendor P Coated Polycarbonate
             (3 Yr. EMMA) 1.0 min. Rain
             Exposure.   100x

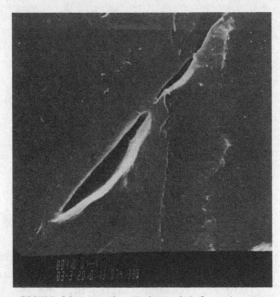

FIGURE 14.   Vendor P Coated Polycarbonate
             (3 Yr. EMMA) 1.0 min. Rain
             Exposure.   3000x

164

SLURRY ABRASION/EROSION BEHAVIOR OF METAL-CERAMIC COATINGS

Alberto A. Sagüés*
Gordon A. Sargent+
Diane K. Spencer*

*Institute for Mining and Minerals Research
University of Kentucky
Box 13015, Iron Works Pike
Lexington, KY 40512

+Department of Metallurgical Engineering and Materials Science
University of Notre Dame
Box E
Notre Dame, IN 46556

Abstract: Four different abrasion- and corrosion- resistant, metal-ceramic coatings
and a carbon steel control alloy have been investigated with a wet rubber wheel
abrasion tester, and with a slurry impingement erosion tester. The results show
that the abrasion resistance of these materials is directly proportional to their
hardness and varies over a wide range. In contrast, the resistance to slurry erosion varied less significantly from material to material, and did not correlate with
their hardness ranking.

The results are discussed in terms of the applicability of each type of test.

Key words: Abrasion; coatings; erosion; fossil fuels; slurries.

INTRODUCTION:

Advanced fossil fuel technologies, such as coal conversion, oil shale retorting, tar
sand processing and others, require handling of slurries in erosive and abrasive
conditions. The slurries can be water or organic medium based, with particles of
coal, coal ash, sand and minerals, or mixtures thereof. Process components must
handle the slurries at wide combinations of pressure, temperature and flow velocity.
The slurries sometimes contain corrosive components, which may act synergistically
with the mechanical action of the solid particles in creating severe wear
situations. Materials deterioration under these circumstances has been observed
frequently in pilot plants used in developing the technologies. Reference (1) is a
DOE newsletter containing numerous examples of materials and component failures
during the handling of fossil fuel related slurries. These occurrences have indicated the need for understanding the mechanisms of damage and for predicting the
performance of materials in the handling of particulates. As a result, the
Institute for Mining and Minerals Research (IMMR) is conducting failure analyses and
exposures of test coupons in pilot plants, as well as laboratory studies where
slurry erosion and abrasion are characterized.

Coatings consisting of metallic binders with ceramic second phase inclusions are
used in critical process components to improve their corrosion and mechanical wear

165

performance. As part of the IMMR's effort, four coatings covering a range of applications were selected to study their performance and modes of deterioration in abrasive and erosive slurry environments.

RESULTS AND DISCUSSION

The compositions, properties, and typical service applications of these coatings are listed in Table 1. The metallic substrate for the coatings was in all cases 1018 carbon steel. The same steel was also used as a reference standard material for comparison purposes. The hardness of the coatings was measured in Diamond Pyramid Hardness (DDH) units, using an indentor force of 300 g. The hardness of the 1018 steel reference blocks was measured to be ∿64 HRB, and was converted to equivalent DPH units (∿110 DPH) for the Table listing. The coated surfaces were ∿5.6 cm X 2.5 cm (2.2 in X 1 in), and the substrate thickness was 1.25 cm ($1/2$ in). Some of the erosion test samples were cut to smaller sizes to improve weighing accuracy. The coatings were supplied by the Linde Division of Union Carbide Corp., of Indianapolis. One of the coatings, LW-26A, is proprietary and its composition cannot be disclosed at this time. Additional information on the properties and applications of some of the other coatings can be found in References (2) to (4).

Two types of test were performed: A wet sand rubber wheel abrasion test (WSAT), and a slurry jet impingement erosion test.

The WSAT test machine has a 17.5 cm (7 in) diameter wheel, whose outer 1 cm (0.4 in) rim is neoprene rubber. The wheel rotates submerged in a slurry of silica sand in water. The piece to be tested for abrasion resistance is pressed with known force (222 N, or 50 lb) against the rubber wheel in a "brake shoe" arrangement. Sand particles become trapped between the rubber wheel and the test piece, with resulting wear of the piece. The weight of the piece before and after exposure in the test rig for a predetermined number of wheel revolutions is measured, and the weight loss is determined. The exposed samples were dried in an oven at 50° for 30 minutes before weighing, to remove water that may have been trapped in pores. A run-in of 1000 revolutions, using a wheel whose rim has a hardness of 50 Shore A Durometer, is conducted at the beginning of the test. Three tests, also consisting of 1000 revolutions each, are performed then with wheels of increasing nominal hardness (50, 60, and 70 Shore A Durometer). The resulting weight losses are then plotted in the form of log wt. loss versus actual wheel hardness. A straight line is fitted to the data using a least squares approximation, and the ordinate intersect of that line at 60 Durometer is used to define the result of the test. The WSAT test is the subject of a SAE Recommended Practice being developed (5) and it has been discussed in detail in the literature (6,7). The test parameters used in this study are listed in Table 2. The SAE recommended practice was generally followed. An exception was that the number of revolutions for the LCO-22 coating was 100 instead of 1000, because the wear rate was high and wastage of the full thickness of that coating would have taken place during a much longer test. The sand used in the test was examined with a Scanning Electron Microscope (SEM), and it had the appearance shown in Figure 1. The size and morphology of the sand particles were consistent with the specifications.

The results of the abrasion tests are shown in Figure 2. As stated earlier the weight loss of the LCO-22 has been extrapolated linearly to 1000 revolutions from the 100 revolutions tests; all the other results are from 1000 revolutions tests. Weight losses observed varied from ∿ 500 mg for the LCO-22 down to less than 1 mg

166

for the LW-26A, a range of more than two orders of magnitude. The relative dispersion of the data became greater as the weight loss was smaller. Since the test
pieces weighed near 150 g and had total surface areas exceeding 20 cm^2, it is to be
expected that weighing inaccuracies in the order of one mg would result, even after
careful cleaning and drying procedures. These inaccuracies probably account for a
significant portion of the scatter at the lower end of weight losses.

The results of the abrasion tests are shown in a different fashion in Figure 3.
Here the volume loss (which is a more descriptive parameter of materials wastage
than the mass loss) at the 60 Durometer intersect has been plotted as a function of
the hardness of the materials tested. The volume loss was computed from the weight
loss and the nominal density of the coatings and steel. The abrasion of the four
coatings correlates well with their hardness; the harder the coating the lower is
its volume loss during the abrasion test. The carbon steel deviates from the mono-
tonic behavior of the coatings, but still its volume loss is much greater than that
of the three hardest coatings.

The erosion tests were conducted using the apparatus shown in Figure 4. A stirred
autoclave with a capacity of 3780 ml (one gallon) is filled with a slurry of 3 wt%
alumina in water and then pressurized to a regulated, predetermined value. A valved
slurry transfer line ends in a 3.2mm (1.8 in) nozzle that faces the material sample
to be tested. The test sample can be tilted using a angle vise so that different
angles of jet impingement are possible. The test is conducted by opening the valve
in the transfer line and letting a predetermined amount of slurry impinge on the
sample. The slurry velocity is computed from the amount of time the valve was open,
the amount of liquid recovered and the dimensions of the nozzle. The test parameters used in this study are shown in Table 3. The experimental arrangement, details
of the technique, and another investigation using the same unit are described in
References (8) and (9).

The material loss due to erosion was originally expressed in grams of material lost
per grams of erodent impinging on the sample during the test. This is a dimen-
sionless parameter commonly used in erosion testing. In this paper, however, the
results will be expressed in sample volume loss per gram of erodent. This is done
to normalize the results for materials of different densities. The results are presented graphically in Figure 5. The logaritmic vertical scale used is the same as
that in Figure 3 for the abrasion test results. The first noteworthy feature is
that the erosion losses span an interval of less than an order of magnitude for both
90° (normal) and 30° impingement angles. This is in contrast to the much larger
variation in response observed during the abrasion tests. Secondly, the erosive
wear is not correlated with coating hardness in the 90° erosion tests. Indeed, the
LW-26A, which was by far the best performer in the abrasion tests, eroded at a rate
closer to that of the carbon steel. At 30° the two hardest coating show lower erosion than the other materials, but by not much more than a factor of 2, in contrast
with the behavior of their abrasion response. The carbon steel and the LCO-22 show
distinctly higher erosion at 30° impingement than at 90°. This behavior is characteristic of ductile materials when tested in gas-driven particle impingement devices
(10), and has also been reported in slurry jet tests (8). For the harder materials
(especially the LW-5) the angular dependence is less marked.

The wear surfaces were examined with the SEM. Figure 6 shows the surface appearance
of the LCO-22 after an abrasion test with a 60 Durometer wheel (left, vertical
direction of motion) and after a 30° erosion test (right, liquid motion from top to
bottom). Notice the broad scars and smearing resulting from abrasion, and the indi-

vidual craters produced during the particle erosion. The erosion tracks are by no means all aligned with the overall liquid motion. This behavior is indicative of liquid turbulence and it has been reported elsewhere (8). The LW-26 corresponds to the other extreme of abrasion behavior, and its surface microstructures are shown (using the same format as for the LCO-22) in Figure 7. The abraded surface shows much finer scars and a smoother appearance than that of the LCO-22. The eroded surface shows similar erosion tracks. Both surfaces present features associated with the presence of the carbides in the coating.

The size of the smears and abrasion scars on the LCO-22 corresponds roughly to that of angularities on the surface of the silica sand particles. It should be kept in mind that at the magnification used the average diameter of the sand particles is several times the size of the picture frame used in Figures 6 and 7. In the case of the LW-26A the scars are even finer. Since some of the carbides show signs of having cracked, it is not unlikely that part of the damage may have been caused by carbide fragments.

The erosion tracks on the LCO-22 surface have a size which is in agreement with the angularity of the 400-grit alumina particles used in the test. The particles themselves have a diameter of about 1/4 the size of the picture frame. The nature of the damage is similar to that observed on carbon steel in a similar investigation (8). On the LW-26A the material removal by erosion is likely to take place mainly at the binder, with the carbides sticking out of the surface. This seems to be supported by the appearance of the damage in the picture.

The hardness of the particles used to wear the coatings may have had an important effect in determining the result of the abrasion tests. The silica particles have a hardness of ~800 DPH, which is less than the average hardness of the LW-5 and the LW-26A. As illustrated by Farmer (11), abrasive wear tends to be very low when the hardness of the abrading particles is lower than that of the material being abraded. As a result, the abrasion performance of the hardest coatings may have been much worse had alumina particles (DPH~2000) been used. Similarly, the relative performance of the coatings in the erosion tests might have been different if softer eroding particles had been used (12). Experiments aimed to ellucidate these points are currently in progress.

CONCLUSIONS

The result of this investigation underline the importance of selecting the appropriate kind of test when attempting to select materials based on laboratory experiments. The wet sand rubber test shows abrasive wear rates that are strongly dependent on the hardness of the coating. The softest materials wore several hundred times faster than the hardest coating. The reproducibility of the WSAT results is best for the materials tested that showed the most abrasion. The abrasion of the hardest coating may have been due in part to particles other than sand; i.e., fragments of broken or dislodged carbides.

The erosive wear response is very different from that observed in the abrasion tests. There is much less dependence on the hardness of the coating. The cobalt-based coatings and carbon steel tended to show a "ductile" erosion behavior. The WC coatings showed less dependence of erosion on the angle of particle impingement.

The relative hardnesses of abrading, eroding particles and the test materials may have played an important part in the difference observed between the two types of tests. Further studies will include determining the effect of particle hardness and morphology on the rates of both modes of wear.

ACKNOWLEDGMENTS

The authors are indebted to Godwin Oghide and Donald Parris for having conducted the abrasion and erosion tests, respectively. The cooperation of M. K. Keshavan and R. C. Tucker of Union Carbide, Linde Division in providing the coating samples along with helpful discussions and encouragement is greatly appreciated. Jill Heink and Gerry Estes took the SEM micrographs. V. K. Sethi has reviewed the manuscript.

The work reported herein was supported by the Kentucky Energy Cabinet (KEC) through a grant from the University of Kentucky Institute for Mining and Minerals Research.

REFERENCES

1) Materials and Components in Fossil Energy Applications - A DOE Newsletter DOE/FE-0053/43, April 1983, and previous issues.

2) J. M. Quets and R. C. Tucker, Thin Solid Films, 84 (1981) 107.

3) T. A. Taylor, J. Vac. Sci. and Tech., 12 (1975) 790.

4) R. C. Tucker, T. A. Taylor and M. H. Weatherly, "Plasma deposited Mcraly Airfoil and Zirconia/MCrAly Thermal Barrier Coatings" presented at the 3rd Conf. on Gas Turbine Materials in a Marine Environment. Bath, England, 20 to 23 Sept., 1976.

5) Wet-Sand Rubber-Wheel Abrasion Test Method. Society of Automotive Engineers.

6) H. S. Avery "An analysis of the rubber wheel abrasion test", in Wear of Materials 1981, Proc. of International Conference on Wear of Materials, March 30-April 1981, American Society Mechanical Engineers, New York, (1981) p. 367.

7) P. A. Swanson and R. W. Klann "Abrasive wear studies using the wet-sand and dry sand rubber wheel tests", Ibid, p. 379.

8) G. A. Sargent, D. K. Spencer and A. A. Sagues, "Slurry Erosion of Materials" in Corrosion-Erosion-Wear of Materials in Emerging Fossil Energy Systems, Proceedings of the January 27-29, 1982 Conference in Berkeley, CA. Al Levy, Editor, National Assoc. of Corrosion Engs., Houston, (1982) p. 196.

9) D. K. Spencer "Erosion of 1050 Steel by water based slurries" M.S. Thesis, University of Kentucky, Lexington, KY (1982).

10) I. Finnie in Source Book on Wear Control Technologies, American Society for Metals (1978) p. 220.

11) H. N. Farmer "Factors Affecting Selection and Performance of Hard Facing Alloys" in Proc. Symposium on Materials for the Mining Industry,

Vail, CO, July 1974. R. Q. Barr, Editor, Publ. by Climax Molybdemum Company, Greenwich, CT, (1975) p. 85.

12) Al Levy and P. Chik "The Effect of Erodent Composition and Shape on the Erosion of Steel" LBL-15242, submitted to <u>Wear</u> (1982).

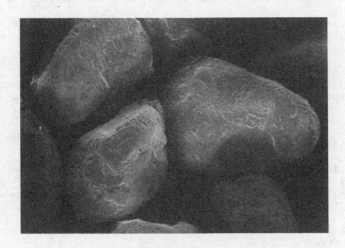

100μm

Figure 1. Appearance of the silica sand used in the abrasion tests.

Table 1.

MATERIALS TESTED

Material	Method of Application	Composition	Uses	Approximate D.P. Hardness	Minimum Coating Thickness (Mils)
LCO-17	D-Gun, Heat Treated	Co-25Cr-7Al-10Ta-0.75Y-2C-0.7Si + 10% Al_2O_3	High Temperature Wear Resistance, Corrosion Resistance	740	8
LCO-22	Plasma, Heat Treated	Co-32Ni-21Cr-8Al-0.5Y	High Temperature Corrosion Resistance,	490	8
LW-5	D-Gun Not Heat Treated	WC-20Cr-7Ni	Wear Resistance, Slurries, Petroleum	1080	10
LW-26A	Fused Coating, Heated Treated	WC + Binder (Proprietary)	--	1150	10
1018 CS Standard	--	--	--	110 (HRB 64)	--

Table 2.

ABRASION TEST PARAMETERS

Apparatus: Fargo Abrasion Testing Machine
                                    SAE Model (wet abrasion)

Wheel Diameter: 177 mm (6.97") typical

Load   220 N (50 lb) Nominal

Liquid: Distilled Water, 940 c.c.

Abrasive: A.F.S. Testing Sand 50-70
                                    Ottawa Silica Co.,
                                    1500g

Number of Revolutions   As per SAE
    and Run In: Recommended Practice, except
                                    for LCO-22, when 100 revolutions
                                    were used and loss multiplied x 10

Table 3.

EROSION TEST PARAMETERS

Nozzle Diameter: 3.2 mm (1/8")

Nozzle to Sample Distance: 6.3 mm (1/4")

Jet Velocity: 18 m/s (54FPS)

Eroding Particulates: 400 Grit Alumina

Slurry Composition: 3 wt% Alumina in water

Solids Loading: 500g (five 100g runs)

Impingement Angles: 30° and 90°

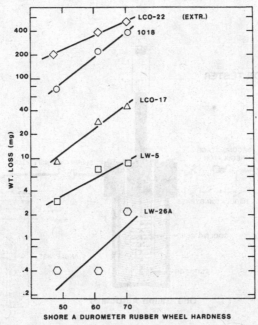

Figure 2. Weight loss versus Durometer reading for abrasion tests of four coatings and the base steel. All tests are for 1000 turns, except for LCO-22, where the result has been extrapolated from 100 turns.

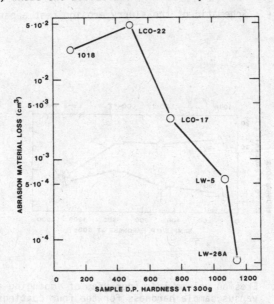

Figure 3. Abrasion volume loss versus sample hardness for the four coatings and the base steel.

173

JET EROSION TESTER

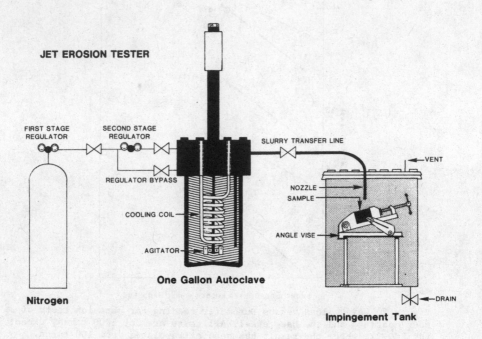

Figure 4.   Schematic of the slurry erosion test apparatus.

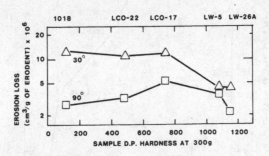

Figure 5.   Erosion volume loss (per gram of impinging erodent)
versus sample hardness for the four coatings and
the base steel.

EROSION TEST, 30 DEG

50 μM

ABRASION TEST

Figure 6. SEM appearance of the wear surface of the LCO-22 coating.

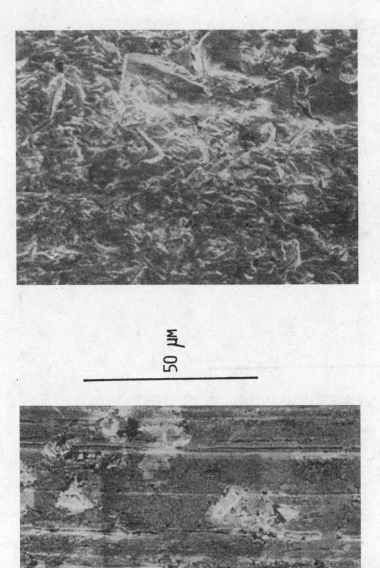

EROSION TEST, 30 DEG

ABRASION TEST

50 μM

Figure 7. SEM appearance of the wear surface of the LW-26A coating.

TRIBOLOGICAL PROPERTIES OF MAGNETRON SPUTTERED TITANIUM NITRIDE

S. Ramalingam
University of Minnesota
Minneapolis, Minnesota 55455

Ward O. Winer
Georgia Institute of Technology
Atlanta, Georgia 30332

Abstract: Magnetron sputtering of titanium in an argon-nitrogen ambient allows the
deposition of well-bonded, thin coats of stoichiometric TiN on a variety of ferrous
and non-ferrous substrates. Materials protected with such hard coats can exhibit
good wear resistance under a variety of tribological contact conditions. To demonstrate that the expected wear resistance can be obtained in practice, sliding and
rolling contact friction and wear tests have been carried out on a number of coated
samples. The test results obtained are presented in this work.

Introduction: In a tribo-contact, surface coatings may be used to lower friction
and/or wear. Generally, soft coatings are used for the former. Hard coats are
appropriate for the latter. In the case of soft coats, an attempt is made to confine
the frictional dissipation in a thin, low shear strength film interposed between the
tribo-elements. Molybdenum disulfide coatings and silver plating are two examples.
Although friction can be lowered, the benefits obtained with soft coats persist only
as long as the film deposited remains on the tribological surface. Solid lubricant
soft films are eventually displaced or worn away in heavily loaded tribo-contacts.
To obtain continued benefit from soft coats, a source of material replenishment is
needed. This is often not feasible. Few of the soft coat materials possess shear
resistance as low as those possessed by liquid lubricants. Hence, large reductions
in frictional dissipation are not possible with soft coats. These considerations
suggest that the largest contribution coating technology can make in tribology is
through hard coats.

Wear is a complex process. It may be defined as a process in which progressive loss
of materials occurs as a consequence of mechanical action. Four principal mechanisms
are responsible for material loss in wear. They are: adhesive wear, abrasive wear,
tribo-chemical (corrosive) wear, and surface (fatigue) wear.

In adhesive wear, metal-to-metal contact is accompanied by welding at a finite number
of discrete points of real contact, followed by detachment of material from welded
junctions. According to adhesive wear thoory, the wear rate W, defined as worn volume
per unit sliding distance, is proportional to the load L and inversely proportional to
the yield pressure P_o of the wearing material, with

$$W = k (L / 3P_o)$$

Here, k is a proportionality constant. A linear relationship is observed in wear
experiments provided the nominal contact pressure does not exceed H/3, where H is the
bulk hardness of the wearing material. Modifications of the adhesive wear model are
made to take the friction coefficient into account. Two additional constants, to
account for surface contamination and to describe material behavior under stress
loading, are used. However, wear volume is still an inverse function of the yield

pressure. Since yield pressure P_o and hardness H are related through a constant C, raising hardness of contacting bodies offers a means of lowering adhesive wear even when corrections are introduced to account for the effect of friction.

Abrasive wear models also lead to wear equations where the wear volume is inversely proportional to the yield pressure. Hence, in both adhesive and abrasive wear, wear volume per unit sliding distance is adequately described by a general equation of the form:

$$W = k* (1 / H)$$

where k* is a constant. It is therefore seen that elementary models of adhesive and abrasive wear suggest that hard coatings can afford protection against adhesive and abrasive wear.

Hardness levels necessary to obtain wear resistance are readily estimated. By examining abrasive wear as a function of the hardness ratio H_a/H_m, where H_a is the hardness of the abrading species and H_m that of the abraded material, Krushchov [1] reports little or no wear in the hardness ratio range of 0.72 to 1.12. Based on Richardson's data [2], it has been suggested that hardness ratios of the order of 0.4 are sufficient to obtain negligible wear [3]. To provide wear protection in sliding against steel in its fully hardened condition with hardnesses of the order of 800 to 1000 HV, counterface hardnesses in the Vickers range of 1500 to 2000 are hence seen to be sufficient. A number of hard components meet this requirement. They are listed in Table 1 with corresponding microhardness characteristics in Table 2.

Tribo-chemical or corrosive wear is a consequence of interaction with the ambient atmosphere. Temperature rise accompanying energy dissipation in tribological contacts enhances the rate of interaction. Milder forms of tribo-chemical wear in air are known to be oxidative [4]. To lower the tribo-chemical wear rate, the rate of interaction with the environment will have to be lowered. In contrast to metals, thermodynamically stable compounds of metals exhibit much lower rates of oxidation. This is due to the large negative free energy of formation of most compounds of metals. Use of hard, stable compound coatings can thus confer protection against adhesive wear, abrasive wear, and tribo-chemical wear.

Surface or fatigue wear is usually not a serious problem in sliding contacts. In heavily loaded, hertzian contacts where fatigue wear is often a problem, failure initiation almost always occurs at a finite depth below the surface and breaks through to the surface as a result of flaw growth. Surface coating is, hence, unlikely to offer protection against fatigue wear. As surface coatings are known to raise the scuffing load in sliding contacts, we may regard its use as generally beneficial even though it may not afford specific protection against classical fatigue wear.

Hard Coat Thickness and Bond Strength Required for Wear Protection: The surface hardness that can be obtained with hard coating is much greater than that obtained with traditional surface hardening processes. With hard coats, the H_a/H_m ratio relevant to abrasive wear is lowered easily to values in the 0.2 to 0 6 range where negligible wear occurs. Coatings comparable in thickness to those produced by traditional surface hardening are then not necessary for wear protection, since wear of hard coated surfaces is expected to be negligibly small. Experience with hard coats under severe unlubricated sliding conditions as in metal machining indicates that coatings as thin as 1 to 10 μm are sufficient to afford wear protection. Similar results are reported following 4-ball wear tests [5]. Thin coats, a few μm in thickness, should therefore be satisfactory.

The use of thin hard coats limits the possible temperature gradient and the magnitude of thermal stresses at the coating-to-substrate interface. Thin layers of hard materials, visualized as Griffith solids, often exhibit higher fracture stresses and elongations with lower coating thicknesses. They should, therefore, behave as compliant coatings. Wear studies involving titanium coatings on elastically and plastically soft substrates at nominal contact stresses of the order of the yield stress of the substrate (6) lend support to this view. Since a sound theoretical basis for the choice of hard coat thickness does not exist and empirical experience suggests that hard coats, a few µm in thickness ought to be satisfactory, thin films less than 10 µm in thickness were used in this study.

Several coating techniques are available for low temperature coating. They may be suitable for wear protection only if satisfactory substrate-to-coating bond strengths can be obtained with these processes. The suitability of a hard coating process can only be established after an estimate of the bond strength requirements is made. Bond strength requirement estimates are made by examining the contact stress and conditions of contact.

In a typical contact with lubrication, the surfaces may be completely separated from each other by an intervening fluid film or the surfaces may be in intermittent contact with significant asperity interaction. In hydrodynamic contacts, the maximum pressure rarely exceeds 75 MPa and the coefficient of friction lies in the 0.001 to 0.01 range. Maximum Hertz pressures of 0.5 to 3.0 GPa and traction coefficients of 0.02 to 0.12 are typical in full film elastohydrodynamically lubricated contacts. In mixed film EHL, the Hertz pressure range is frequently the same as in full film EHL, but the associated traction coefficients range between 0.02 and 0.3. The shear stresses experienced at the coating-to-substrate interface are calculated for these cases with the contact pressure and the friction or traction coefficient. The needed interface shear strengths in the worst case are of the order of 60 to 900 MPa, larger values being those for mixed film EHL.

The use of unlubricated sliding contacts is not common. A friction contact can be technically described as unlubricated when sliding contact assemblies are made with self-lubricating materials. The mean contact pressures in unlubricated systems usually range between 3 and 70 MPa. Typical values are of the order of 15 MPa. Maximum coating-to-substrate bond strength requirements in such cases, when lubricant failure is total, will rarely exceed 70 MPa. A bond strength requirement of 70 MPa for such instances is realistic.

Bond strength requirements are thus seen to cover a broad range. Interface shear strengths of the order of 1 GPa are needed in rolling contact bearings, and levels to 2 GPa may be necessary in tooling applications. The requirements in other unlubricated conditions are less stringent.

Summary of Hard Coat Requirements: In summary, the most useful tribological coatings will necessarily have to be hard. Exceptional surface hardness levels are not necessary to assure extended life. Small wear rates obtained with hard coats indicate that well-bonded thin coats are sufficient to obtain wear protection. When several materials that meet the hardness requirements are available, the material exhibiting the lowest friction coefficient is preferred. The coating material that most closely matches the E-alpha product of the substrate, where E is the elastic modulus and alpha is the thermal coefficient of expansion, is preferred. All other conditions being comparable, in a choice among several materials, the most stable of the hard compounds is the preferred choice.

Selection of Hard Coating Process: Many processes are available to coat tribological components with wear-resistant layers. The choice, in the present context, is restricted to those capable of producing thin coats. The justification is as follows: use of thin coatings eliminates the need for refinishing either to meet surface finish requirements or for dimension control; thin coats require smaller coating times; they also lower the demand for scarce strategic materials.

The coating process selected must meet specific needs based on operational requirements. Coatings produced must consistently possess adequate coating-to-substrate bond strength. Since the coatings are compounds and compound properties are strong functions of stoichiometry, the process selected should be able to produce coatings with perfect or near-perfect stoichiometry. Coating thickness and uniformity control must be possible. Significant microstructure and property changes must not accompany coating. Coatings produced must be dense and, as far as possible, fine grained and equiaxed. Process reproducibility, ease of coating control, and coating cost considerations are other standard requirements that any industrially viable process must meet.

Coating Technology Used: To deposit thin films of hard metal compounds, chemical vapor deposition (CVD) and physical vapor deposition (PVD) techniques may be used. Since CVD requires high temperatures (800°C and more), a PVD process, high rate magnetron sputtering, was used in this study. Magnetron sputtering employs a well-designed electron trap to enhance ionization and, hence, the feasible thin film deposition rate.

Description of the Coating System and the Coating Conditions Used: The coating system used consists of four principal subsystems. The subsystems of the coating plant are: a vacuum plant, a high rate sputtering head, a sputtering atmosphere control system, and a power supply. It is designed to facilitate metallic coating at high rates as well as coating of hard material compounds such as TiN. The latter is obtained by reactive sputtering.

The vacuum plant used is a commercial CVC 14" vacuum system suitably modified for the present needs. A vacuum chamber defined by a 12 inch diameter by 12 inch high aluminum cylinder, a top plate, and a base plate is used. The vacuum chamber can be evacuated to pressures less than 10^{-5} torr with a 4 inch diffusion pump and a 5.6 CFM mechanical pump. The system is equipped with a 4 inch gate valve and roughing valves for vacuum sequencing and to control ambient pressure during sputtering. The system is equipped with a bleed valve to bring the vacuum chamber to atmospheric pressure for specimen loading and unloading. Dynamic pumping practice was employed to sweep contaminants and desorbed species.

The top plate of the vacuum chamber carries the sputtering head and is electrically insulated from it with the insulator ring. It also carries a sputtering pressure monitoring gage, a shutter control, and a gas inlet. The base plate of the system contains a number of electrical and instrumentation feed-throughs.

To produce golden yellow TiN by magnetron sputtering in the present system, it is necessary to operate the system at high cathode currents, low pressures, and high voltages. In the course of this work, it has been found that the best TiN films are produced with the present system at the following operating conditions:

System voltage:	925 to 950 V
Cathode:	1000 to 1200 ma
System pressure:	1.75 to 2.25 millitorr
Substrate-cathode distance:	4.6 to 6 cm
Argon flow rate:	2 SCCM
Nitrogen flow rate:	2.5 SCCM
Magnetron field strength on the cathode surface (central pole):	1200 Gauss
Cathode diameter:	5 cm

Typically, a coating run is for 30 to 60 minutes. Following the coating, nitrogen flow is cut off, and the system is left to cool down for several minutes. After shutting off the argon flow and isolating the vacuum chamber from the pumping system, air is bled into the system and the top plate is removed to take the coated specimens out of the coating system.

Principal heating during coating is due to the condensation of the TiN on the substrate surface. There is some secondary heating due to electron leak from the vicinity of the central pole piece. Measurement of specimen surface temperature immediately following the cessation of coating showed that surface temperatures did not exceed 375°F.

Characterization of the Coatings Produced: Micro-hardness measurements and x-ray diffraction techniques were used to characterize coatings deposited. Micro-hardness measurements on taper-sectioned test samples showed that coatings with hardnesses greater than HV 2000 are routinely produced by reactive sputtering of titanium. Titanium nitride, a defect compound, can possess a range of stoichiometries. Distinct colors are associated with specific ranges of stoichiometry, golden yellow being that of the perfect compound. This was obtained routinely.

Since, in sputtering, non-equilibrium structures are usually produced and the films deposited are under stress, x-ray analysis is not an unambiguous indicator of film stoichiometry. Additional x-ray analysis problems are introduced by preferred texture in coatings. Despite these difficulties, x-ray analysis shows that the films produced are titanium nitride compounds. Expected x-ray peaks and those observed are in agreement. Supporting evidence is provided by micro-hardness and color data.

Tribological Evaluation of Coatings Produced: In engineering practice, the tribological contact conditions encountered span a broad range. High contact stress, low speed sliding contacts as well as high speed, low stress contacts are common. More severe contact conditions are encountered in Hertzian contacts. In this instance, lubrication is generally provided to generate a hydrodynamically-lubricated bearing system. There are, in addition, a number of EHL systems with varying slide-to-roll ratios.

It is neither possible nor feasible to evaluate coatings produced at each combination of contact stress, relative velocity and contact type, i.e., sliding, rolling, and rolling with varying slide-roll ratio, etc,, encountered in practice. Three types of representative tests were hence chosen and used to evaluate the coatings produced. Two of these are sliding contact tests, LFW-1 and LFW-6 tests at high and low contact stress, respectively. The rolling contact tests carried out used a special rig which allowed the slide-roll ratio to be varied during the tests.

181

A Brief Description of Tests Used: To assess tribological properties in high stress, low speed tests, a Faville-LaValley alpha model LFW-1 test machine was used. The test machine was calibrated following ASTM test procedures (7), and the tests themselves were conducted with a procedure similar to that used for calibration. All tests were carried out at a sliding speed of 0.13 ms^{-1} with an unformulated paraffin-based mineral oil of viscosity 26 x 10^{-6} m^{2}s^{-1} at a bath temperature of 43°C and AISI 52100 test rings hardened to 60 RC with an r.m.s. surface finish of 0.4 microns. The tests were 5000 revolutions of test ring in duration with a sliding distance of 550 m. The test loads used were selected to yield an initial Hertzian contact stress of 60% of the estimated yield strength of the substrate material. From calculations and observations of friction and wear, the elastohydrodynamic lubrication load-carrying component in the film was found to be insignificant. A new ring and fresh oil were used in each LFW-1 test.

For low contact stress, high speed tests, a sliding thrust washer assembly similar to that used in Faville-LaValley LFW-6 tribo tester was employed. The test sample configuration is that of an annulus rotating on a flat. Since early tests indicated that the coatings can withstand high PV products, a contact pressure of 500 psi (3.4 MPa) and a test speed of 1800 rpm (PV product = 250,000 psi-ftmin^{-1} = 9.5 MPa - m.s^{-1}) were used. Either n-hexadecane or an unformulated mineral oil, same as that used in LFW-1 tests, was used. With either of these fluids, test samples of fully hardened T-15 high speed steel and M-50 HSS failed instantaneously under the test conditions used. Tests lasting five minutes at the specified test conditions were taken to indicate successful coating. Sample bulk temperatures during tests were monitored with a type K thermocouple and temperature rise of about 100°C is common in five minutes of tests.

A specially constructed concentrated contact simulator was used to assess traction characteristics in rolling contact tests. The test system used is described in a previous publication (8). Typically, AISI 52100 discs and rollers were used with TiN coating to assess traction characteristics as a function of slide-roll ratio. Some test samples with duplex coatings were also used (a soft coat overlay on TiN coats). Test procedures are described elsewhere (8).

Results of High Contact Stress, Low Speed Sliding Contact Tests: Since low carbon steels, cast irons, and aluminum alloys are among materials commonly used for machinery construction, test samples of mild steel (SAE 1018), a gray cast iron (grade 20), and wrought aluminum alloys (2024 and 6061) were TiN coated and tested. LFW-1 tests were also carried out on two additional non-ferrous alloys, a cast magnesium alloy and Ti-6Al-4V alloy, to determine the utility of TiN coating for reduction of wear of these alloys. Both these materials are prone to severe galling in normal sliding contact.

The specific test conditions used and the test results obtained from LFW-1 tests are shown in Tables 3 to 6. In every instance, hardened AISI 52100 steel rings were used as friction counterfaces. As may be seen from the data presented, thin coats of TiN afford wear rate reductions of approximately two orders of magnitude. It is seen that films as small as 2 μm in thickness are sufficient to bestow significant wear protection.

To further illustrate the significant wear rate reduction obtained with thin films of TiN, the profiles of the wear scars produced during LFW-1 tests were recorded. Some representative data are reproduced in Figure 1. Unambiguous and striking improvement in wear resistance due to the presence of thin layers of TiN is clearly evident. Representative SEM micrographs of coated LFW-1 test samples following wear tests are presented in Figure 2. Although there is evidence of film cracking just outside the wear scar in the region stressed in tension, Figure 2, it is clear that there is little large scale film failure and consequent loss of wear protection.

Test results from Faville-6 sliding contact tests are summarized in Table 7. While severe surface distress and large increases in friction were found to occur immediately following contact in tests with fully hardened but uncoated high speed steel samples, TiN coated samples of ferrous alloys and non-ferrous alloys survived the full five minute test duration. The tests were terminated following five minutes of contact solely because of the significant temperature rise that accompanies prolonged, loaded contact at high speeds (500 ft/min:2.5 m/s). It should be recognized that neither n-hexadecane nor the unformulated mineral oil used in the test are "lubricants" under the test conditions used. Moreover, the test conditions selected are particularly severe with PV products of 250,000 psi - ft.min^{-1} (9.5 MPa$\cdot$m's^{-1}). It is noted that the design limit for the bronze with the highest PV rating, aluminum bronze, is of the order of 50,000 psi - ft.min^{-1} or 2 MPa$\cdot$m's^{-1}.

In the 500 psi, 500 ft.min^{-1}, Faville-6 tests, both the contacting surfaces are coated. Thus, the contact is between TiN and TiN. Friction and wear processes that accompany the Faville-6 tests are those between two "ceramics" or refractory material compounds. Severe wear is not expected, and the results may be compared with alumina-on-alumina sliding contact tests where very little wear is usually found to occur. The substantial free energy of formation, -66.1 kCal/mole (9), of titanium nitride affords significant thermodynamic stability to the coated body. Corrosive wear can, therefore, be expected only at moderate contact temperatures (of the order of 1000-1200°F or more). The contact surface is, hence, only expected to exhibit "burnishing" as a consequence of severe sliding contact. The SEM micrographs obtained, Figure 3, show this to be the case.

In the case of cast irons containing flake graphite, adhesion between the TiN film and the graphite phase can be a problem. The SEM micrographs from both the LFW-1 and the Faville-6 tests, not surprisingly, do show small scale debonding at graphite flakes. Despite this, test results were satisfactory. In passing, it is noted that whenever nodular cast irons with massive graphite nodules are hard-coated, there is likely to be large scale film failure. Hard coating of such materials is not preferred.

Representative test results obtained from roller-on-disc traction tests are presented in Figure 4. When TiN coated rollers are tested against TiN coated discs, although the contact stress (mean Hertzian stress) was 1 GPa and the traction coefficient in excess of 0.2 at a slide-roll ratio greater than 0.04, film failure is not observed. The test result suggests that bonding strengths, i.e., film-to-substrate adhesion strengths, will have to be in excess of 0.2 GPa for film survival. Well-bonded films are thus seen to be possible even on hardened steels. Concentrated contact tests were made for TiN coated (1.9μm) cast iron, AISI 1018, and AISI 4340. The conditions were: peak Hertz pressure 620 MPa, pure rolling at 2 m/s, and mineral oil at 43°C. Microscopic debonding was observed. With cast iron, major debonding was at the graphite flakes as shown in Figure 5. For the steels, debonding was the result of substrate collapse as shown in Figure 6.

To assess if TiN coatings are feasible for use in piston ring applications where normally sprayed coatings of chromium, molybdenum, etc., are used, preliminary coating trials have been carried out. Automotive rings of ductile iron have been coated with TiN using the magnetron reactive sputtering technique. Twist bend tests for film adhesion and fatigue tests carried out on coated rings show that films deposited do not bond well to the graphite nodules. Film debonding is initiated at these specific sites as may be seen from Figure 7. Fatigue tests, on 109.22 mm diameter piston rings with 4.470 mm x 3.124 mm cross section coated with 1 μm of TiN, also show film debonding at graphite nodule sites (50 mm fatigue displacement, unidirectional strain, 6.5 cycles/min). Observation was after 9700 cycles. It would, hence, appear that TiN coating for rings will require substrate material change. Since piston rings normally operate at contact pressures of the order of 50 psi, it should be possible to use steel rings with TiN coatings in such applications (expected PV product is 125,000 psi-ft.min^{-1}; system with a 3 inch stroke, operating at 5000 rpm is assumed).

In the case of adiabatic engines with ceramic cylinder walls, hard coated rings used in conjunction with synthetic oils containing additive packages appear to hold promise. One may wish to use alumina or zirconia sputter coated rings (depending on the wall ceramic) in preference to TiN coated rings. Coating technology would appear to hold considerable promise in this application. (Note: Preliminary coatings of TiN have been produced successfully on Inconel alloy substrates.) Nitride coated, refractory superalloy rings may permit operation of adiabatic engines constructed of solid or sprayed silicon nitride cylinders with minimal lubrication if the accompanying frictional losses are tolerable.

Solid lubricant coatings can be beneficial for friction reduction and to assist run-in process. To assess frictional and durability characteristics of solid lubricant coatings, Faville-6 thrust washer tests were carried out. Test samples were coated with a thin film of a WeSe$_2$-In-Ga alloy (commercial designation Westinghouse Compact). Since solid lubricant films will eventually wear away in the absence of a source for replenishment, tests were carried out with a duplex coating to protect the surface following loss of soft coat, i.e., with a hard coat and a soft coat overlay. Test results obtained are presented in Figure 8. Both tests were carried out at a nominal contact pressure of 90 psi at 70 rpm.

It can be seen that duplex coated test samples in pure sliding yield frictional tractions as low as 0.06. This friction coefficient is comparable to that obtainable with (nearly) full hydrodynamic lubrication at low speeds. as may be seen from Figure 8B, satisfactory durability is attainable, at least at low contact speeds.

Results obtained in rolling contact traction tests with MoS$_2$ and WSe$_2$-In-Ga alloy are shown in Figure 9. It is recognized that tungsten diselenide is a good solid lubricant, comparable to well-established MoS$_2$ coatings. Scanning electron micrographic studies carried out show that sputtered coatings are well bonded and are not easily worn away as long as the slide-roll ratio is close to or equal to zero. SEM studies also show that at slide-roll ratios greater than 0.04, there can be substantial film loss.

The results obtained suggest that in machinery applications, soft coats are suitable provided that in sliding contacts, large life and durability are not required and contact stresses are low. These are the conditions experienced during run-in at the piston ring-cylinder wall interface. It would thus appear that solid lubricant films are worthy of more detailed studies for this application. In Hertzian contacts, soft coat films are useful, if slide-roll ratio is close to zero. This condition is experienced in many mechanical systems and in the automotive drive train. Further tests and evaluation are needed to assess suitability of soft coats in such applications.

It should be noted that in all cases where cast iron is the preferred material of construction, film coating techniques as practiced here do not appear satisfactory since

184

film-graphite bond-strength is much poorer than that for film-metal bonding. It should also be recognized that film coating techniques are also not appropriate to offset problems in Hertzian contact, if substrates are not hardened adequately.

Summary: Tribological characteristics of thin hard coats laid on a variety of ferrous and non-ferrous substrates have been tested. The thin hard coats used were titanium nitride films deposited by reactive magnetron sputtering of metallic titanium. High contact stress, low speed tests show wear rate reductions of one or more magnitude, even with films a few micrometers thick. Low contact stress, high speed tests carried out under rather severe test conditions show that thin films of TiN afford significant friction and wear protection. Thin hard coats are also suitable to improve the friction and wear performance of rolling contacts. Satisfactory film-to-substrate bond strengths can be obtained with reactive magnetron sputtering.

Acknowledgements: The authors thank S. Bair, Y. Shimazaki, and B.-Y. Ting for experimental assistance.

References:

1. Kruschov, M.M., Wear, 28, 1974, p. 69.
2. Richardson, R.C.D., Wear, 11, 1968, p. 245.
3. Halling, J., Principles of Tribology, MacMillan & Co., London, 1975.
4. Quinn, T.F.J., "A Review of Oxidational Wear," NASA Contractor Report 3686, June 1983.
5. Scott, D., Wear, 48, 1978, p. 283.
6. Ramalingam, S. and W.O. Winer, Thin Solid Films, 73, 1980, p. 267.
7. ASTM Standard D2714-68, Am. Soc. Test. Mat., Philadelphia, PA, 1968 (reapproved 1978).
8. Ramalingam, S., W.O. Winer, and S. Bair, Thin Solid Films, 84, 1981, p. 272.
9. Weast, R.C. (Editor), Handbook of Chem. & Phys., 52nd Edition, Chem. Rubber Co., Cleveland, OH, 1971, p. D-71.

TABLE 1. HARD COAT MATERIALS FOR TRIBOLOGICAL APPLICATIONS

Carbides	Nitrides	Oxides	Borides
TiC	TiN	TiO_x	TiB_2
HfC	HfN	HfO_2	HfB_2
ZrC	ZrN	ZrO_2	ZrB_2
TaC	TaN	Ta_2O_5	TaB_2
VC	VN	V_2O_3	VB
NbC	NbN	--	NbB_2
Cr_3C_2; Cr_7C_3; $Cr_{23}C_6$	CrN	Cr_2O_3	--
SiC	Si_3N_4	SiO_2	--
WC; W_2C	--	--	WB
Mo_2C	--	--	MoB
--	AlN	Al_2O_3	--
TiC-TiN	TiN-TiC	$TiC-Ti_xO_y$	--
TiC-VC	(Ti,V)N	--	--
Ti-Si-C	--	--	--
$(Fe,Mn)_3C(b)$	--	--	--
--	(Si,Al)N	--	--
--	Fe_4N	--	--

All materials listed have been deposited as thin coats with one or
another of the coating technologies.

Table 2: MICROHARDNESS OF TYPICAL HARD COAT MATERIALS

VICKERS HARDNESS, kg/mm^2

Element	Carbide	Nitride	Boride
Boron	3700	--	--
Chromium	1600-Cr$_7$C$_3$	2200	1800
	1300-Cr$_3$C$_2$	1083-CrN	
Hafnium	2270-2650	1640	2250-2900
Molybdenum	1800-MoC	--	2350
Niobium	2400-2850	1396-NbN	2100-2400
		1720-Nb$_2$N	
Silicon	3500	--	--
Tantalum	1800-2450	1220	2450-2910
Titanium	2000-3200	1200-2000	2200-3500
Tungsten	2100-2400	--	2400-2660
	1450-W$_2$C		
Vanadium	2460-3150	1520-1900	2070-2800
Zirconium	2360-2600	1150	2250-2600

Note: Literature microhardness values span a moderately wide range. A single specific value is usually not representative. Transition metal oxides, nitrides, and carbides can vary widely in stoichiometry and are mutually soluble. Variations in hardness reported are due to variations in stoichiometry and purity. Most borides, especially the hexagonal borides, are highly anisotropic.

TABLE 3: SUMMARY OF LFW-1 TESTS ON MILD STEEL BLOCKS

Block: AISI 1018 Mild Steel, Anneated.

Ring: AISI 52100 Hardened Steel, R_C = 62

	COATING THICKNESS (microns)			
	None	2.0	3.5	4.2
Load (kN m^{-1})	92.3	92.3	92.3	92.3
Hertz Pressure (MN m^{-2})	436.5	436.5	436.5	436.5
Coefficient of Friction[a]	.20/.16	.16/.14	.15/.15	.14/.13
Block Mass Loss (mg)	9.2	0.02	0.05	0.14
Ring Mass Loss (mg)	0.1	1.0	1.23	1.08
Block Wear Coefficient[b] K(x10^{-6})	32.5	0.26	0.64	1.79
Ring Wear Coefficient K(x10^{-6})	1.34	2.68	3.30	2.90
Sliding Distance (m)	110	550	550	550

---

[a] The coefficient of friction values quoted are the time-averaged values for the beginning and end of the tests.

[b] For coated specimens, a Vickers hardness of 2200 kgmm^{-2} for TiN was used.

TABLE 4: SUMMARY OF LFW-1 TESTS ON CAST IRON BLOCKS

Block: Grade 20 Cast Iron, As Received
Ring: AISI 52100 Hardened Steel, $R_c = 62$

	COATING THICKNESS (microns)		
	None	2.0	4.2
LOAD (kN m^{-1})	122.7	122.7	122.7
HERTZ PRESSURE (MN m^{-2})	362.6	362.6	362.6
COEFFICIENT OF FRICTION [a]	.21/.25	.15/.12	.18/.15
BLOCK MASS LOSS (mg)	7.89	0.15	0.12
RING MASS LOSS [b] (mg)	-0.20	2.07	3.04
BLOCK WEAR COEFFICIENT [c] K(x10^{-6})	129.2	1.45	1.15
RING WEAR COEFFICIENT K(x10^{-6})	--	4.18	6.15
SLIDING DISTANCE (m)	16	550	550

[a] The coefficient of friction values quoted are the time-averaged values for the beginning and end of the tests.

[b] A negative loss indicates that the ring gained mass.

[c] For coated specimens, a Vickers hardness of 2200 kgmm^{-2} for TiN was used.

TABLE 5: SUMMARY OF LFW-2 TESTS ON ALUMINUM

Block: Aluminum 6061-T651 Treatment
Ring: AISI 52100 Hardened Steel, R_c = 62

	COATING THICKNESS (microns)				
	None	2.1	2.5	3.8	4.2
LOAD (kN m^{-1})	63.0	63.0	63.0	63.0	63.0
HERTZ PRESSURE (MN m^{-2})	257.0	257.0	257.0	257.0	257.0
COEFFICIENT OF FRICTION [a]	.14/.16	.14/.13	.18/.14	.17/.13	.19/.13
BLOCK MASS LOSS (mg)	2.77	1.64	1.09	0.16	0.05
RING MASS LOSS (mg)	1.00	0.40	0.75	0.85	1.84
BLOCK WEAR COEFFICIENT [b] K(x10^{-6})	24.56	-- [c]	19.98	3.01	0.94
RING WEAR COEFFICIENT K(x10^{-6})	19.65	3.15	2.95	3.34	7.24
SLIDING DISTANCE (m)	110	275	550	550	550

---

[a] The coefficient of friction values quoted are the time-averaged values for the beginning and end of the tests.

[b] For coated specimens, a Vickers hardness of 2200 kgmm^{-2} for TiN was used.

[c] Mass loss is sum of TiN and Aluminum. Wear coefficient not calculated.

190

TABLE 6: SUMMARY OF LFW-1 TESTS ON ALUMINUM ALLOY

Block: 2024 Aluminum Alloy, T351 Condition
Ring: AISI 52100 Steel, Hardened to R_C 62

	UNCOATED	TiN COATED		
		1.0 M	3.6 M	4.5 M
LOAD (kN m^{-1})	63.0	63.0	63.0	63.0
HERTZ PRESSURE (MN m^{-2})	257.2	257.2	257.2	257.2
COEFFICIENT OF FRICTION[a]	.16/.18	.15/.16	.19/.16	.18/.16
BLOCK MASS LOSS (mg)	30	14.9	0.08	0.04
RING MASS LOSS (mg)[b]	-0.44	0.89	1.84	2.52
BLOCK WEAR COEFFICIENT K[c] ($\times 10^{-6}$)	102	--[d]	1.56	0.73
RING WEAR COEFFICIENT K($\times 10^{-6}$)	--	3.57	7.38	10.1
SLIDING DISTANCE (m)	330	550	530	550

---

[a] The coefficient of friction values quoted are the time-averaged values for the beginning and end of the tests.

[b] A negative loss indicates that the ring gained mass.

[c] For coated specimens, a Vickers hardness of 2200 kgmm^{-2} for TiN was used.

[d] Mass loss is sum of TiN and Aluminum. Wear coefficient not calculated.

TABLE 7: SUMMARY OF TEST RESULTS LFW-6, THRUST WASHER TEST RESULTS

Contact Pressure: 500 psi

Rotational Speed: 1,800 rpm (1.062" dia.)

PV Product: 250,000 psi-ft/min

Results:

Uncoated	Steel	M50	N2 Min.Oil[3]	Inst.Fail
	Steel	M50	N2 (400 psi)	Inst. Fail
	Steel	M50	Motor Oil[4]	5* Min.
TiN-Coated[2]	Aluminum	6061[1]	n-hexa dec.	9.5* Min.
	Aluminum	6061[1]	n-hexa dec.	7.5 Min.
	Mild Steel (AISI 1018)		Motor Oil	5* Min.
	Mild Steel (AISI 1018)		N-2 Oil	5* Min.
	Mild Steel (AISI 1018)		Mineral Oil[5]	5* Min.
	Cast Iron		Motor Oil	5* Min.
	Cast Iron		N-2 Oil	5* Min.
	Cast Iron		Mineral Oil	5* Min.

1   Test terminated without failure.

2   Aluminum Sample No. 1 TiN Coating: 2.7 microns
    Aluminum Sample No. 2 TiN Coating: 4.2 microns
    Mild Steel and Cast Iron TiN Coating: 1.9 microns

3   N2 Min. Oil is R620-15 with 4.0% Polymer

4   SAE 10W-40 HD

5   Unformulated Paraffinic Mineral Oil R620-15

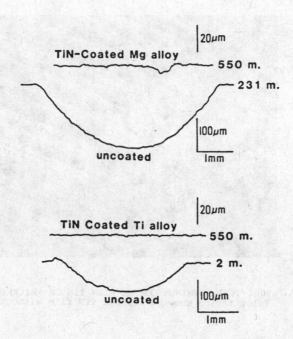

FIGURE 1: WEAR SCARS PRODUCED ON TEST BLOCKS WITHOUT AND WITH TiN COATINGS

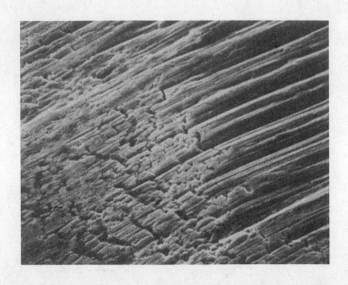

FIGURE 2A: 6061-T651 ALUMINUM COATED WITH TiN OF THICKNESS 2.5μm
Inlet region showing cracked TiN film without debonding (900x)

Figure 2B: Cast Iron
Inlet region showing cracked TiN film without debonding (2 kx)

Figure 3A: Aluminum coated with 2.0 μm TiN and run with n-hexadecane, edge of contact region (240x)

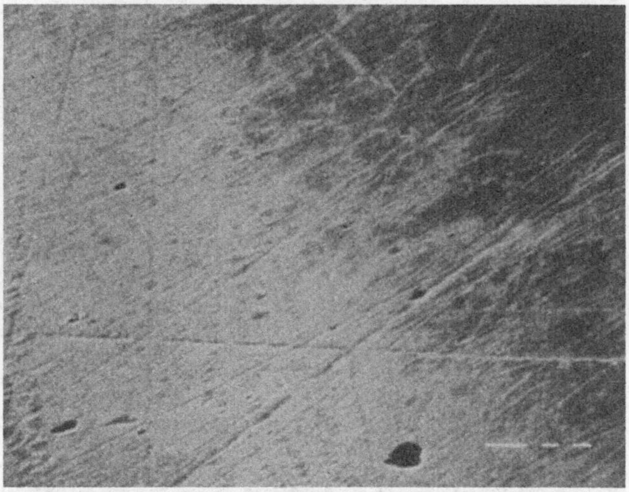

Figure 3B: Mild steel coated with 1.9 μm TiN and run with N2 oil, edge of contact (740x)

FIGURE 3:   SEM MICROS OF LFW-6 TEST SPECIMENS
See text for test conditions.

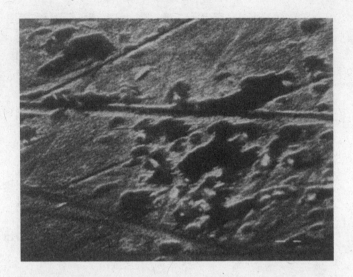

Figure 3C: Mild steel coated with 1.9 μm TiN and run with motor oil, in wear track (3.4kx)

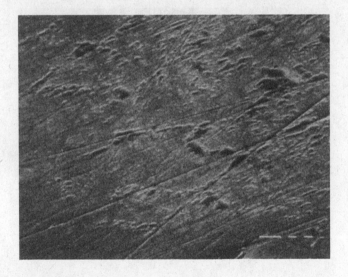

Figure 3D: Cast iron coated with 1.9 μm TiN and run with motor oil, edge of contact region (760x)

196

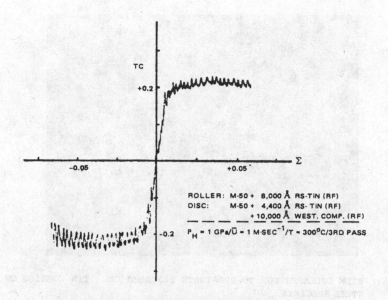

ROLLER: M-50 + 8,000 Å RS-TiN (RF)
DISC: M-50 + 4,400 Å RS-TiN (RF)
————— + 10,000 Å WEST. COMP. (RF)
$P_H = 1\ GPa/\bar{U} = 1\ M\cdot SEC^{-1}/T = 300^\circ C/3RD\ PASS$

FIGURE 4: TRACTION VERSUS SLIDE-ROLL RATIO RECORDED IN A ROLLING CONTACT TEST
OF TiN COATED, HARDENED STEEL SPECIMENS

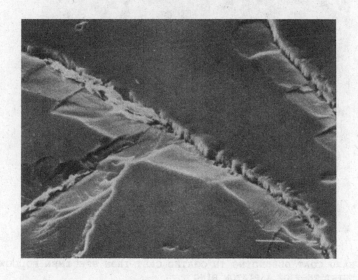

FIGURE 5: FILM DEBONDING IN A CAST IRON AT LOCATION WITH GRAPHITE

FIGURE 6: FILM COLLAPSE DUE TO SUBSTRATE DEFORMATION.   TiN COATING ON A SOFT
STEEL SPECIMEN

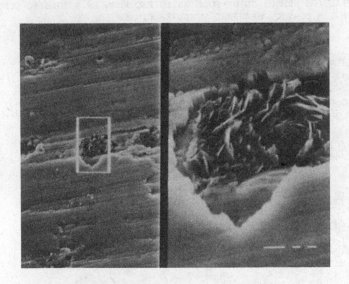

FIGURE 7: HARD COAT DEBONDING IN COATED CAST IRON SPECIMEN FOLLOWING THE
TWIST TEST OF A PISTON RING

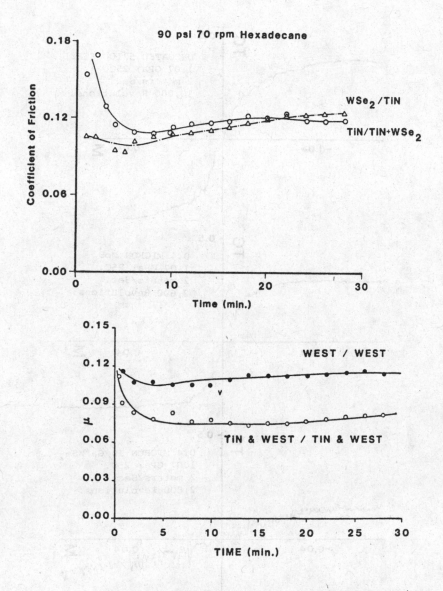

COATED AISI 52100 TEST SAMPLES (HARDENED)

FIGURE 8: FRICTION-TIME TRACE OBTAINED IN LFW-6 TESTS WITH WSe$_2$-In-Ga THIN COATS

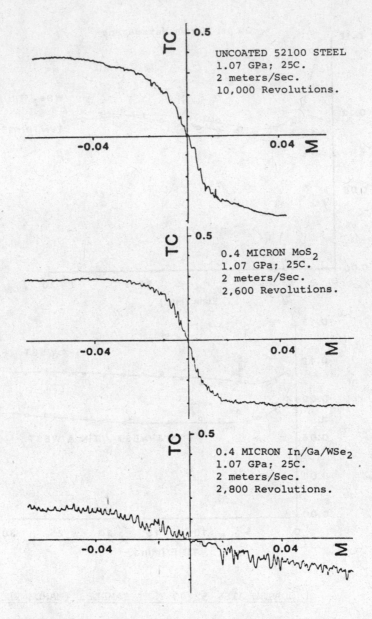

FIGURE 9: TRACTION VERSUS SLIDE-ROLL RATIO WITH MoS$_2$ AND WSe$_2$-In-Ga ALLOY COATINGS

POLISHING WEAR STUDIES OF COATING MATERIALS

Robert N. Bolster and Irwin L. Singer
Naval Research Laboratory
Washington, DC 20375

Abstract: Relative abrasive wear measurements using 1- to 15-μm abrasives have
been found useful in investigating the tribological properties of modified
surfaces and thin films. Modified and reference surfaces were polished
simultaneously using a vibratory polisher or a modified polishing wheel, and
relative wear rates determined from their mass losses. Depth resolution on the
order of 10 nm was readily attained. Wear rates responded to changes in polishing
pressure and speed, bulk hardness, and work-hardening effects. A series of carbon
steel surfaces of varied hardness gave expected wear rates with a 15-μm abrasive.
With 1- and 3-μm diamond, the wear resistances of the softer surfaces were
anomalously high. Work-hardenable alloys such as 304 stainless steel showed
increased wear resistance relative to carbon steel when polished at low wear
rates. A major application of this technique has been the investigation of ion-
implanted metal surfaces, where the modified region usually was only a few hundred
nm deep. Wear-resisting effects of N^+ and Ti^+ on steel and C^+ on Ti alloy have
been measured. The increased wear found when 304 stainless steel was implanted
with certain species led to an understanding of the microstructural changes
produced by implantation. The properties of thin vapor-deposited films and laser-
treated composite surfaces have also been studied.

Key words: Abrasive wear; hardness; ion implantation; polishing wear; thin films;
tribology; wear; work hardening.

Introduction: In studying the tribological properties of thin films, especially
the extremely shallow regions affected by ion implantation, measurements of their
mechanical properties such as hardness are difficult to obtain. Microhardness
measurements using conventional diamond indenters sample depths of 0.2 to 2 μm,
while implanted layers are typically 0.1 μm deep. Hardness has been correlated
with abrasion resistance for a variety of metals and alloys (1,2), usually using
large abrasive particles which penetrated several micrometers into the surfaces.
Rabinowicz developed a relative wear measuring technique using a lapping machine
(3,4) which was effective with finer abrasives. We have adopted this technique
and extended it into the polishing regime by using micrometer-size abrasives, thus
reducing the depth of the abrasive action to a few nanometers.

Experimental Techniques: Absolute abrasive wear rates depend on numerous
variables, some of which are difficult to control. Measuring the relative wear of
two or more surfaces abraded simultaneously under the same conditions produced
much more consistent data. Specimens were always abraded in groups of three,
typically two reference surfaces and one modified or coated one. The specimens
were disks 12.7 mm in diameter and 2.8 mm thick, ground and fine ground as for
metallographic examination, then polished with the machine and abrasive to be used
in the wear determination.

Two polishing machines have been used in these wear studies. A vibratory polisher, providing relatively low wear rates, was equipped with three holders of equal mass, each with three recesses for specimen disks, as shown schematically in Fig. 1. For higher wear rates and more control over speed and pressure a wheel polisher equipped with a motor-driven epicyclic specimen carrier was used. Three specimen holders were made, each with three recesses. The holders were in two parts to allow loading the specimens from above, and with adjustable pivots to allow the use of specimens of differing thicknesses. The carrier loading device was modified to allow the holders to be individually weight-loaded. The wheel was run at low speeds; 20, 40, or 90 rpm, and the carrier at 1.4 or 14 rpm. The polishing speed varied with rotation of the carrier, but instantaneous speeds were calculated and integrated to yield mean speeds of 11 to 50 cm/s for the most commonly used combinations.

The laps used on both polishers were of nonwoven synthetic textile. Nominal 3-μm diamond paste (actual range 1 to 5 μm) was the most commonly used abrasive, with 30- and 1-μm diamond and 15-μm emery used in a few cases to study the effect of abrasive size on wear. The abrasives were extended with paraffin oil, and the atmosphere was air dried to a frost point of 200K. The laps were used for numerous runs, with abrasives added daily to make up for losses.

Wear depths were determined from mass losses and the specimen densities. Relative wear resistance (RWR) was calculated by dividing the mean depth change of the reference specimens by the depth change of the modified one in the same holder. To achieve the desired depth resolution of a few nm, it was necessary to weigh the specimen disks with a microbalance to a precision of a few micrograms. Reproducibility of better than 5 μg was normally attained, which corresponded to depths of about 5 nm for steels, so wear depth changes of 10 nm were readily resolved. Changes in wear resistance with depth were less well resolved, spreading to approximately 20 to 30 nm with the 3-μm diamond, because the individual cuts produced by the diamond particles were almost this deep (5).

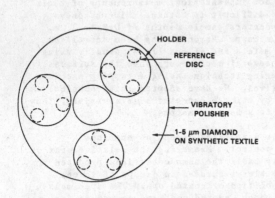

Fig. 1   Abrasive wear apparatus

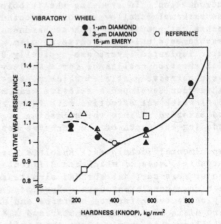

Fig. 2   Wear resistance vs. hardness of a carbon steel under various conditions of abrasion. Solid line: 80-μm abrasive (6).

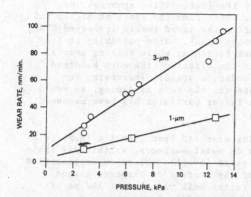

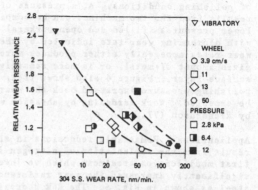

Fig. 3 Effect of pressure on the wear rate of carbon steel (KHN = 400 kg/mm²) abraded with 1- and 3-μm diamond at 11 cms on the wheel polisher.

Fig. 4 The effect of absolute wear rate on the wear resistance of 304 stainless steel relative to carbon steel (KHN = 400 kg/mm²) polished with 3-μm diamond with different machines, speeds, and pressures.

Characteristics of the Polishing Wear: A series of specimens of low-alloy carbon steel (Fe-1C-0.25 Mn-0.2Si), heat-treated to various hardnesses, were used to explore the relationship between hardness and wear resistance with these small abrasives. The results for both polishing machines and several abrasive sizes are shown in Fig. 2. The solid curve is from a study of abrasive wear of a similar steel by fixed 80-μm abrasive by Mutton and Watson (6). The results with specimens of KHN = 400 kg/mm² hardness or higher (measured with a 2-kg load) fit very well, as did the result of the 15-μm emery abrasion of a soft surface. However, unexpectedly high wear resistance was found with the 1- and 3-μm diamond on the softer surfaces. The reason for this is unclear, but it may be due to a change in the mode of wear produced by large and small abrasives. Therefore, the relationship between wear resistance and bulk hardness must be applied cautiously when small abrasives are used.

The load on the specimen holders, and thus the polishing pressure, could be readily varied when using the wheel. Figure 3 shows the effect of pressure on the absolute wear rate of carbon steel (KHN = 400 kg/mm²) for 1- and 3-μm diamond. The wear rates were found to increase linearly with pressure, and were higher with the larger abrasive. The vibratory polisher, with a pressure of 6.4 kPa, gave wear rates of 2 to 12 nm/min. with 3-μm diamond. When a new lap was put into use, the wear rate was found to rise to a maximum in 5 to 10 hours of use, and then decline to a steady state in 30 to 40 hours.

With certain alloys such as 304 stainless steel and Ti-6Al-4V, the wear resistance relative to carbon steel was found to vary with pressure and speed, producing a disparity between the results with the two polishers. Although the speed of the vibratory polisher could not be determined, a tenuous link between the two machines could be established through the absolute wear rate as shown in Fig. 4. Here the wear resistance of the 304 steel relative to carbon steel (KHN = 400 kg/mm²) has been plotted against its absolute wear rate resulting from a variety of polishing conditions. At a pressure of 6.4 kPa (half-filled symbols) the results from the two machines can be compared. Data from the wheel at higher and lower pressures (filled and open markers) confirm the trend toward increased RWR with decreasing wear rate indicated by the broken lines. Slow polishing at low wear rates apparently either produces more work hardening while wearing away a given amount of metal, or is more sensitive to the effect of the work hardened surface layer. Figure 4 also shows that, at moderate speeds, increasing the polishing pressure increased both the wear rate and the work hardening, as would be expected. Work hardening by abrasion with larger particles has been measured by Richardson (7).

Applications: Remarkable reductions in sliding wear had been reported (8,9) to result from the implantation of nitrogen ions in metal surfaces, so this was the first modification treatment which we investigated (10,11). Nitrogen implantation significantly increased the wear resistance of the surface of tempered carbon steel as shown in Fig. 5. The RWR decreased to the bulk value after 100 nm of wear, commensurate with the nitrogen concentration profile found by Auger analysis. The same treatment was found to provide no additional wear resistance in fully hardened steels (5).

Implantation of nitrogen in 304 stainless steel was unexpectedly found to reduce the RWR to about half of the bulk value, in apparent contradiction to its beneficial effect on sliding wear. The work hardening effects previously described provided the clue needed to understand this phenomenon. In Fig. 5 are shown the wear resistances vs. depth of three N-implanted steels (304, the reference carbon steel, and an unhardened tool steel) relative to carbon steel. It is presumed that the work hardened surface layer normally present on the 304 and M-7 steels was altered after implantation with nitrogen, causing them to wear more like the carbon steel. As the implanted layer was worn through, the work hardening reappeared. This was confirmed in further work with Vardiman (12) who showed that martensite produced by rolling deformation or abrasion of the normally austenitic 304 steel was transformed back to the softer austenite by nitrogen implantation. Implanted nickel has been found to produce the same effects in 304 steel.

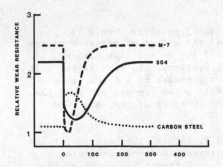

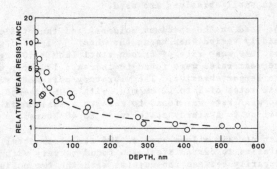

Fig. 5 Wear resistances vs. depth for several steels implanted with nitrogen ions, relative to carbon steel.

Fig. 6 Relative wear resistance of titanium-implanted 304 stainless steel vs. depth (11).

Implantation of titanium in steels was found to produce extremely wear resistant surfaces. Figure 6 shows that the surface of 304 stainless steel was found to be 10 times as wear resistant as the bulk after the implantation of 4.6 x 10¹⁷ Ti ions/cm² (11). This dose produced a maximum concentration of 30 atomic percent at a depth of 70 nm. The same dose in hardened 52100 bearing steel produced the effect shown in Fig. 7 (5). Further investigation of this effect (13,14) has shown that carburization of the implanted layer, which occurred during the implantation process, was essential to the formation of the wear resistant layer, which also reduced friction and wear under dry sliding conditions.

The formation of titanium carbide by the implantation of carbon into Ti-6A1-4V alloy is possibly responsible for the high RWR shown in Fig. 8. This treatment has also been found to increase the fatigue life of titanium by 4 to 5 times at high stress levels (15).

Films such as those produced by electroplating, chemical vapor deposition, vacuum evaporation, etc., can also be examined. The very limited depth of the abrasive action reduces spurious effects due to the substrate or the interface. Figure 9 shows the results of some wear experiments with niobium nitride films on sapphire substrates (16). These films were prepared under controlled conditions to provide surfaces of known composition and structure, so that these parameters could be related to their tribological properties. Nb_2N films were more wear resistant than NbN films and showed reorientation of the crystal structure after abrasion. The dip in RWR at 500 nm resulted from a change in stoichiometry to $Nb_{1.2}N$, which was caused by a temporary rise in the nitrogen pressure during film growth. Films V had more lattice distortion than films IV, and higher Knoop hardness.

Composite surfaces prepared by injecting carbide powders into laser-melted aluminum and titanium alloys have also been examined by this technique (17). Results obtained with 3-μm diamond at 23 cm/s are shown in Fig. 10. As would be expected, higher carbide concentrations provided greater wear resistance, and the titanium matrix was superior to aluminum. In titanium, the tungsten carbide was found to be considerably less effective than the titanium carbide. Dissolution and subsequent precipitation of the TiC occurred during melting and cooling, and may have helped to increase the wear resistance. With 30-μm diamond (not shown) the wear rates were lower, the relative wear resistances of the titanium composites were higher, and less difference was found between the different carbides.

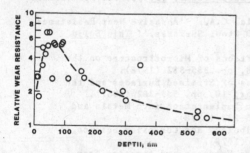

Fig. 7 Relative wear resistance of titanium-implanted 52100 bearing steel vs. depth (5).

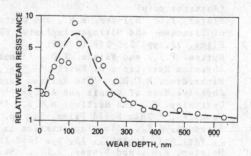

Fig. 8 Relative wear resistance of carbon-implanted titanium alloy vs. depth.

205

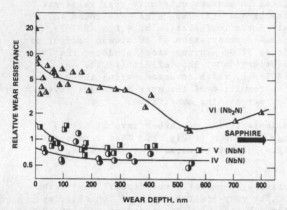

Fig. 9 Wear resistance relative to sapphire of three niobium nitride films vs. depth (16).

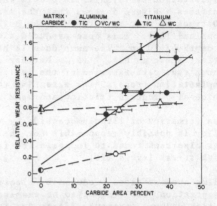

Fig. 10 Relative wear resistance of carbide composite surfaces abraded with 3-μm diamond at 23 cm/s (17).

Conclusions: Abrasive wear in the polishing regime can be done quantitatively to yield relative wear data for very thin surface layers. The wear resistances of metals modified by ion implantation to depths of only 100 nm have been explored using this technique. Surface microstructure and the changes induced by wear and various treatments can be studied, yielding vital information on this region most important to the science of tribology.

References:

1. Khruschov, M.M., ''Principles of Abrasive Wear,'' Wear, 28, pp. 69–88 (1974).

2. Rabinowicz, E., Friction and Wear of Materials, John Wiley and Sons, Inc., New York (1965), pp. 169–176.

3. Rabinowicz, E., ''Abrasive Wear Resistance as a Materials Test,'' Lubr. Engr., 33, 7, pp. 378–381 (1977).

4. Rabinowicz, E., Doherty, P., and Boyd, D.M., ''Measurement of the Abrasive Wear Resistance of Hard Coatings,'' Thin Solid Films, 53, pp. 301–302 (1978). (Abstract only)

5. Singer, I.L., Bolster, R.N., and Carosella, C.A., ''Abrasive Wear Resistance of Titanium– and Nitrogen–Implanted 52100 Steel Surfaces,'' Thin Solid Films, 73, pp. 283–289 (1980).

6. Mutton, P.J., and Watson, J.D., ''Some Effects of Microstructure on the Abrasion Resistance of Metals,'' Wear, 48, pp. 385–382 (1978).

7. Richardson, R.C.D., ''The Maximum Hardness of Strained Surfaces and the Abrasive Wear of Metals and Alloys,'' Wear, 10, pp. 353–382 (1967).

8. Dearnaley, G., and Hartley, N.E.W., ''Ion Implantation into Metals and Carbides,'' Thin Solid Films, 54, pp. 215–232 (1978).

9. Hirvonen, J.K., ''Ion Implantation in Tribology and Corrosion Science,'' J. Vac. Sci. Technol., 15, pp. 1662–1668 (1978).

10. Bolster, R.N., and Singer, I.L., ''Surface Hardness and Abrasive Wear Resistance of Nitrogen–Implanted Steels,'' Appl. Phys. Lett., 36, pp. 208–209 (1980).

11. Bolster, R.N., and Singer, I.L., ''Surface Hardness and Abrasive Wear Resistance of Ion-Implanted Steels,'' <u>ASLE Trans.,</u> 24, pp. 526-532 (1981).
12. Vardiman, R.G., Bolster, R.N., and Singer, I.L., ''The Effect of Nitrogen Implantation on Martensite in 304 Stainless Steel,'' <u>Metastable Materials Formation by Ion Implantation,</u> Elsevier Science Publishing Co., Inc., New York (1982), Eds. Picraux, S.T. and Choyke, W.J.
13. Singer, I.L., Carosella, C.A., and Reed, J.R., ''Friction Behavior of 52100 Steel Modified by Ion Implanted Ti,'' <u>Nucl. Instr. and Meth.,</u> 182/183, pp. 923-932 (1981).
14. Singer, I.L., and Jeffries, R.A., ''Surface Chemistry and Friction Behavior of Ti-Implanted 52100 Steel,'' <u>J. Vac. Sci. Technol.,</u> A1, p. 317 (1983).
15. Vardiman, R.G. and Kant, R.A., ''The Improvement of Fatigue Life in Ti-6Al-4V by Ion Implantation,'' <u>J. Appl. Phys.,</u> 51, pp. 690-694 (1982).
16. Singer, I.L., Bolster, R.N., Wolf, S.A., Skelton, E.F., and Jeffries, R.A., ''Abrasion Resistance, Microhardness, and Microstructures of Single-Phased Niobium Nitride Films,'' <u>Thin Solid Films,</u> to be published.
17. Ayers, J.D. and Bolster, R.N., ''Abrasive Wear with Fine Diamond Particles of Carbide-Containing Al and Ti Alloy Surfaces,'' to be published.

SOLID LUBRICATION OF STEEL BY SbSbS$_4$

L. K. Ives
Metallurgy Division
National Bureau of Standards
Washington, DC 20234

M. B. Peterson
Wear Sciences, Inc.
Arnold, MD 21012

ABSTRACT

The lubricating behavior of the amorphous solid, antimony thioantimonate (SbSbS$_4$), in the form of a dry powder and as compressed pellets is investigated and compared to MoS$_2$ and several other sulfides. The friction and wear response of the dry powders was determined by utilizing a three-pin-on-disk test configuration. Pins were of 52100 steel and disks were of 0-2 tool steel. With SbSbS$_4$ a relatively thick and tenacious film was produced on the surface. The coefficient of friction was high ($\sim$0.7) and the wear rate of the steel specimen surfaces was also high. By means of electron microscopy analyses it was determined that FeS$_2$ was a constituent of the film on worn surfaces. Thus, it was concluded that chemical reaction between SbSbS$_4$ and steel during sliding was an important factor in the friction and wear processes.

Sliding experiments with compressed pellets of SbSbS$_4$, MoS$_2$, Sb$_2$S$_3$, FeS$_2$, and Fe$_{0.9}$S were used to study the friction, film forming, shear, and adhesion characteristics of these solid materials in the absence of metal to metal contact. A pin-on-ring configuration was employed with 52100 steel rings. The lubrication mechanism of SbSbS$_4$ is discussed on the basis of the results of these experiments. Simple models of solid film lubrication are presented to assist in the analysis.

Key words: Antimony thioantimonate; compressed pellets; friction; MoS$_2$; pin-on-disk test; SbSbS$_4$; solid lubrication; three-pin-on-disk test; wear.

INTRODUCTION

The successful application of a solid lubricant, as with any lubricant, requires a detailed knowledge of the operating system parameters, including contact stresses, temperature, chemical nature of the environment, and length of service, together with a corresponding detailed knowledge of lubricant properties. It is especially important to recognize that different solid lubricants may not function by the same mechanisms. One of the challenges of solid lubricant research is to determine the nature of the lubrication mechanisms and thereby provide the means for predicting performance under a given set of system operating conditions. The purpose of the present investigation is to determine the solid lubrication behavior and associated mechanisms of antimony thioantimonate.

The compound antimony thioantimonate (SbSbS$_4$) is one of several complex sulfides of Sb and As which were discovered to produce a substantial increase in the extreme pressure and anti-wear performance of lubricating greases at concentrations of a few percent (1-3). The improvement was significantly greater in most cases than that produced by MoS$_2$ and soluble organic additives to which comparisons were made. Recent studies (4,5) have shown when grease containing 5 wt% SbSbS$_4$ is used to lubricate steel under boundary conditions that not only is a thin film of SbSbS$_4$ formed on the surface but also there is a chemical reaction that results in the formation of iron sulfide. It was therefore hypothesized that SbSbS$_4$ was in this respect similar to other sulfur containing additives which are known to be effective extreme pressure agents.

In this investigation the lubrication properties of $SbSbS_4$ in the form of dry powder and as compressed pellets were studied. Application in both cases was to steel surfaces. A three-pin-on-disk test configuration was employed to study the friction and wear behavior of the dry powders and a pin-on-ring configuration was used for the compressed pellet studies. For comparison, similar experiments were conducted on MoS_2 and several other sulfides of related interest.

EXPERIMENTAL

The powder materials used in this investigation are listed in Table 1. Only $SbSbS_4$ and MoS_2 powders were examined in the three-pin-on-disk test. Included in Table 1 are the chemical formula, commonly used descriptive name, crystal structure, hardness, purity, and particle size. A particle size value is not given for $SbSbS_4$; however, transmission electron microscope examination of several powder samples indicated a range of sizes from a few tenths to larger than a micrometer. Many of the larger particles appeared to be agglomerations of smaller particles.

Pellets were prepared by compressing the powder materials into a spherical die having a radius of 4.76 mm using the device shown schematically in Fig. 1. The die material was 304 stainless steel. A manually operated press was used to apply a pressure of 900 MPa. Among the pellets that were prepared, those of $SbSbS_4$ in particular were found to contain numerous cracks and usually fractured into several pieces when removed from the die. This problem was remedied by applying the pressure in a sequence which consisted of alternately increasing the pressure by an increment of 180 MPa and then releasing it by an increment of 90 MPa until the full 900 MPa was reached. This sequence was reversed on reducing the pressure from 900 MPa. Small cracks were still observed in the surface of the pellet but these did not appear to have an adverse effect on the experiments carried out in this investigation. The hardness of the pellets was

Table 1

	Lubricant	Crystal System	Bulk Hardness (Mohs)	Pellet Hardness KHN	Purity (%)	Particle Size
$SbSbS_4$	Antimony thioantimonate	Amorphous	-	90±7 [a]	-	-
MoS_2	Molybdenum disulfide	Hexagonal	1-1.5	13±4 [b]	98.2	-325 mesh
Sb_2S_3	Stibnite	Rhombohedral	2	45±6 [a]	98	-
FeS_2	Pyrite	Cubic	6-6.5	82±24 [c]	99.9	-100 mesh
$Fe_{0.9}S$	Pyrrhotite	Hexagonal	3.5-4.5	114±35 [a]	99.9	-100 mesh

a) 100 g load
b) 25 g load
c) 500 g load

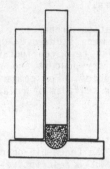

Fig. 1. Schematic drawing of device used to prepare compressed pellets.

measured using the Knoop microhardness method. The results are given in Table 1.
The pellet hardness values obtained are strongly dependent not only on the
intrinsic hardness of the materials themselves but also the powder particle size
characteristics and the extent to which the materials were caused to flow at the
applied pressure. Thus for the hardest material examined, FeS_2, there was probably little plastic flow and, compared to its bulk hardness, the pellet hardness
was relatively small.

Three-Pin-On-Disk Test: The three-pin-on-disk specimen configuration is shown
schematically in Fig. 2. The pins used in this investigation were stock 52100
steel bearing balls having a hardness of ~60 HRC. They were clamped into the
upper holder shown in Fig. 2 so that only sliding motion was obtained. Sliding
was achieved by rotating the pin holder against the stationary lower specimen disk.
The diameter of the path traced out by the pins was 44.5 mm. Guides attached
behind the pins moved powder into the sliding path after it had been pushed aside
by the pins. This feature permitted long term operation without the need to
frequently reapply the powder. The machine also incorporated a load cell which
permitted the measurement of friction force.

The specimen disks were of 0-2 tool steel heat treated to a hardness of 61 HRC.
The surfaces were finished by surface grinding which produced a surface roughness
of 0.2 μm R_a (arithmetic average roughness).

Prior to testing, pin and disk specimens were ultrasonically cleaned in hexane and
then in acetone. The tests were conducted in air at an ambient temperature of
~22°C and a relative humidity that was in the range 30 to 60%.

Pin-On-Ring Test: The pin-on-ring specimen configuration used to study the sliding
behavior of compressed pellet specimens is shown schematically in Fig. 3. The
pellets were cemented to the end of a short brass rod with epoxy. The rod was then
clamped in the test machine. The test rings were of 52100 bearing steel heat
treated to a hardness of 62 ±1 HRC. The rings were of the same design as those
specified in the LFW-1 test (ASTM D2714) (6), having an outside diameter of ~ 35
mm and width of ~8 mm. However, the specified ground finish of 0.1 to 0.2 μm
was not used. Instead, to avoid the possible abrasion of the relatively soft pellets, the rings' surfaces were metallographically polished while turning on the
test machine. The final finishing step employed 1/4 μm grade diamond pol-

Fig. 2. Schematic drawing of three-pin-on-disk test configuration. Bearing ball pins are clamped into upper holder. Guides behind pins return powder to the wear track after it has been pushed away by the pin during sliding.

Load

Fig. 3. Pin-on-ring specimen configuration schematic.

ishing compound resulting in a surface roughness of 0.05 µm R_a measured parallel to the axis of the ring. In addition to avoiding abrasion of the compressed pellets, the relatively smooth surfaces also reduced the mechanical bonding of the pellet material to the ring surface and therefore gave a better indication of chemical adhesion between the pellet material and steel (oxide covered).

All of the pin-on-ring test results on compressed pellet materials reported here were obtained at a load of 18 N and sliding speed of 5.5 cm/s. The test environment was air with an ambient temperature of 22°C and a relative humidity that ranged from 30 to 50%.

RESULTS

Three-Pin-On-Disk Results: A typical trace of coefficient of friction for SbSbS$_4$ powder recorded as a function of sliding distance is shown in Fig. 4. A sliding speed of 8 cm/s was employed. After a short run-in period at 60 N, the full test load of 242 N was applied. The coefficient of friction rose rapidly from an initial value in the range 0.2 to 0.3 at the start of run-in to a steady state value lying between 0.6 to 0.7, achieved soon after the full load had been applied. The friction behavior was characterized by rapid fluctuations and occasional

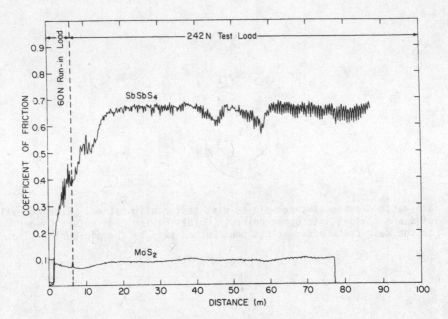

Fig. 4. Coefficient of friction traces for SbSbS$_4$ and MoS$_2$ plotted against sliding distance. An initial run-in load of 60 N was followed by a test load of 242 N. 8 cm/s sliding speed.

larger excursions. For comparison, the coefficient of friction trace for MoS$_2$ powder obtained under the same conditions is also plotted in Fig. 4. As is typical of MoS$_2$ the coefficient of friction is quite low, ~0.1.

Pin and disk contact surfaces were examined after the tests. A smooth and shiny film was found on the pin and disk surfaces when MoS$_2$ was used as the lubricant. With SbSbS$_4$ a film was also present but in this case it was considerably thicker and was not smooth. Photographs of the film on disk and pin surfaces are shown in Fig. 5a and b, respectively. Non-adherent material was removed prior to photographing these surfaces by brief ultrasonic agitation in hexane. Although there are gaps in the film where the steel disk surface is visible, the small size of these regions together with the large thickness of the film indicate that there was little if any contact between the steel surfaces. That is, it appears that there was nearly always an intervening layer of film material present during sliding. Thus, even though the coefficient of friction was approximately equal to that which would be obtained by sliding steel on steel without lubrication, this was not the source of the high friction coefficient observed here.

In addition to a high coefficient of friction, the rate of wear with SbSbS$_4$ was also found to be high. Although an extensive evaluation was not carried out a wear rate of ~1X10^{-4} mm^3/m for the pins was obtained in tests conducted at a load of 242 N and sliding speed of 8 cm/s. The wear volume was calculated from the measured scar diameter on the pins. Perhaps the most interesting finding with respect to wear concerned the topography of the worn steel surfaces. Fig. 6 is a scanning electron micrograph showing the appearance of a worn pin surface after the film had been dissolved in a saturated aqueous solution of NaOH. Dissolution of the film was rapid and there was no noticeable effect on the 52100 steel as was evident on examining areas adjacent to the wear scar. Rather than a

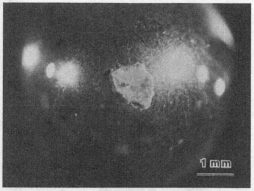

Fig. 5. (a) Optical micrograph of wear track on 0-2 tool steel disk showing thick film produced by SbSbS$_4$ dry powder lubricant. (b) Corresponding 52100 steel pin surface showing attached film.

Fig. 6. Scanning electron micrograph of worn 52100 steel pin surface after dissolving film in NaOH.

pattern of grooves or scratches parallel to the sliding direction, the surface has an etched-like appearance.

Fragments of film were removed from the worn surface and examined by means of transmission electron microscopy. One example of such a fragment is shown in Fig. 7. Small particles are visible within the fragment displayed under bright field conditions in Fig. 7a. These particles are crystalline and produced the diffraction pattern shown in Fig. 7c. A dark field image obtained with reflections from two of the relatively intense inner rings is shown in Fig. 7b. Particles contributing to the selected reflections are strongly visible. Diffraction patterns were analyzed and found to be in good agreement with cubic FeS$_2$ (iron pyrite). Apparently then, sliding promoted a chemical reaction between SbSbS$_4$ and the steel surface with a result that FeS$_2$ was produced. No other reaction product, for example, Sb$_2$S$_3$ or ternary compounds of Fe, S, and Sb were found. It should be noted that faint, broad rings from amorphous SbSbS$_4$ are also present in Fig. 7c. The film fragment therefore consists of a mixture of the original SbSbS$_4$, or an

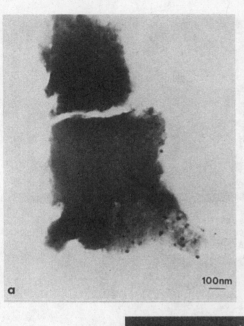
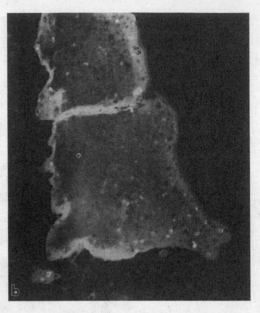

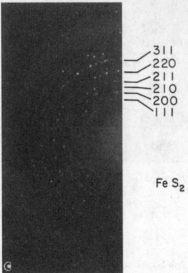

311
220
211
210
200
111

Fe S$_2$

Fig. 7. Transmission electron micrographs of a film fragment from 52100 steel
pin surface lubricated with SbSbS$_4$. (a) Bright field micrograph. Small
particles are visible. (b) Dark field micrograph showing strongly
diffracting crystalline particles. (c) Diffraction pattern from film
fragment. Pattern is indexed as cubic FeS$_2$ (pyrite). Faint, broad rings
from amorphous SbSbS$_4$ are also present.

amorphous derivative, and the FeS_2 reaction product.

Pin-On-Ring Results: Coefficient of friction traces for $SbSbS_4$ and MoS_2 pellets are shown in Fig. 8. In these experiments, the 18 N test load was applied and then sliding was commenced. A constant sliding speed of 5.5 cm/s was reached within approximately one revolution of the 35 mm diameter ring, i.e., approximately 10 cm distance. The behavior of the coefficient of friction was similar to that recorded in the three-pin-on-disk test. The coefficient of friction rose rapidly from an initial value of ~0.3 to above 0.9. With continued sliding a gradual reduction in coefficient of friction occurred until a value of about 0.85 was reached at the end of the test. There was also an accompanying increase in the amplitude of fluctuations. For the MoS_2 pellet the coefficient of friction was again quite low in comparison with $SbSbS_4$ but with a value of 0.25 it is considerably higher than that obtained when MoS_2 was used as a powder lubricant between hard surfaces. A relatively high coefficient of friction with MoS_2 pellets has been shown previously (7,8).

Wear of both the $SbSbS_4$ and MoS_2 pellets was high. A photograph of the two pellets after sliding for a distance of 108 m is shown in Fig. 9. The worn surface in both cases consists of a few slightly raised smooth areas surrounded by rough regions. Contact with the ring surface at the time the tests were terminated was apparently at these smooth areas. Photographs of the corresponding ring surfaces are shown in Fig. 10. A thick transferred layer of MoS_2 is present on the ring after sliding against the MoS_2 pellet. In comparison the transferred film on the ring tested against $SbSbS_4$ is barely visible. Optical microscopy examination revealed very thin patches of transferred $SbSbS_4$ rather than a continuous layer.

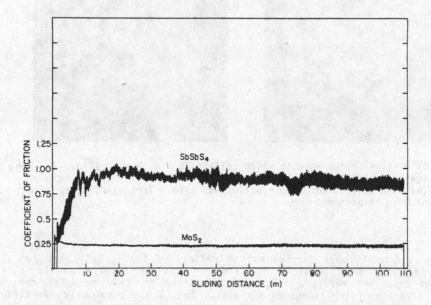

Fig. 8. Coefficient of friction trace for $SbSbS_4$ and MoS_2 pellets sliding on 52100 steel. 18 N load. 5.5 cm/s sliding speed.

215

Fig. 9. (a) SbSbS$_4$ and (b) MoS$_2$ pellet surfaces after sliding for a distance of 108 m.

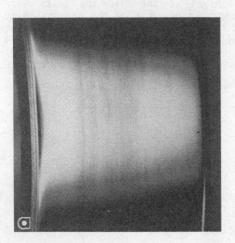 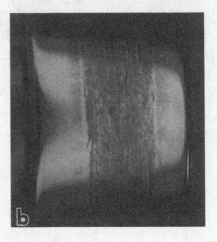

Fig. 10. 52100 steel rings against which pellets in Fig. 9 were slid. (a) Ring against which SbSbS$_4$ was slid. Transferred film is barely visible. (b) Ring against which MoS$_2$ pellet was slid. Irregular, thick film of MoS$_2$ is present.

Pin-on-ring experiments were carried out on pellets of Sb$_2$S$_3$, FeS$_2$, and Fe$_{0.9}$S. These materials represent possible reaction products of SbSbS$_4$ and the steel surface. As described earlier, FeS$_2$ was identified in film material removed from worn three-pin-on-disk specimen surfaces. Fe$_{1-x}$S was previously identified in films on surfaces lubricated with lithium grease containing 5 wt% SbSbS$_4$ (5). Sb$_2$S$_3$, although not identified as a reaction product, is nevertheless the stable sulfide of antimony that can form when SbSbS$_4$ loses a sulfur atom (3). The coefficient of friction traces of Sb$_2$S$_3$, FeS$_2$, and Fe$_{0.9}$S are shown in Fig. 11. For all three materials the coefficient of friction values obtained were less than

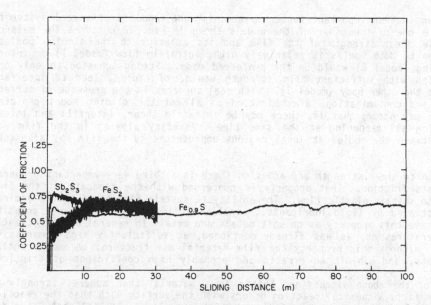

Fig. 11. Coefficient of friction traces for (a) Sb_2S_3, (b) FeS_2 and (c) $Fe_{0.9}S$. 18 N load. 5.5 cm/s sliding speed.

that for $SbSbS_4$. Pellet wear was rapid for Sb_2S_3 and FeS_2 requiring the termination of the test after ~30 m of sliding. In both cases a thick transferred layer was formed on the ring. For $Fe_{0.9}S$ the friction trace showed much smaller fluctuations, the pellet wear rate was quite low and a thin continuous transferred film was formed on the surface.

DISCUSSION

In contrast to the significant improvement in anti-wear and extreme pressure properties realized when $SbSbS_4$ was added to lubricating greases (3,5), its performance as a dry powder lubricant between steel surfaces is poor. The coefficient of friction is high, approximately 0.7, and wear is severe. A thick film was present on the contact surface. Thus, the frictional characteristics are predominantly those of $SbSbS_4$ against steel and do not reflect a significant contribution from steel rubbing steel. This was confirmed by the high coefficient of friction observed when $SbSbS_4$ pellets were slid against steel.

In analyzing the lubricating behavior of $SbSbS_4$ it is convenient to consider simple idealized models of sliding between two bodies with an intervening film. Three different cases can be considered. These are illustrated in Fig. 12a, b and c. In Fig. 12a sliding is considered to occur entirely by intrafilm flow. Bridgman (9) has demonstrated that most solids can exhibit uniform plastic flow behavior at sufficiently high compressive stresses. Fig. 12b depicts a condition where sliding occurs between thin layers of film attached firmly to the two bodies. There is no flow in this case and sliding is entirely interfilm. This mode of sliding is thought to occur with graphite as a lubricant in the presence of water vapor. Hard oxide films at low loads may also exhibit this behavior. The model in Fig. 12c is similar to 12b except that sliding occurs between the film and one or the other of the bodies, that is, interface sliding.

It is clear that an important factor determining the tendency of a real system to behave like one or the other of the models shown in Fig. 12 concerns the balance between the shear strength of the film and its adhesion to the sliding bodies. If adhesion to both bodies is relatively high, intrafilm flow (model 1) or inter-film sliding (model 2) would be the preferred mode. Strong adhesion to only one body coupled with sufficient film strength would, of course, lead to interface sliding at the other body (model 3). With real surfaces in the presence of surface roughness and contamination, a combination of all of the sliding modes depicted in Fig. 12 may occur; that is, there may be intrafilm shear, interfilm and inter-face sliding all happening at the same time. Asperity plowing in the film and contact between the bodies at local regions unprotected by the film may also take place.

In addition to shear strength and adhesion there is a third very important property that affects friction. That property is concerned with the capacity of the film material to sustain plastic flow. The ability to sustain flow is of course admira-bly demonstrated by fluid lubricants. Lamellar solids such as MoS_2 and graphite may also have this property as do soft metals and metals in general at sufficiently high temperatures or, as was already mentioned, at sufficiently high compressive stresses. Without this property the film material may fracture, become irregular in thickness, and exhibit an erratic and probably high coefficient of friction.

In terms of the above discussion, $SbSbS_4$ is a material that adheres strongly to steel. In fact, a chemical reaction occurs with the surface such that the reaction product, FeS_2, is formed during sliding. $SbSbS_4$ is also a material that, as a dry powder, compacts rather readily to form a relatively hard and brittle solid. Pellets of $SbSbS_4$ formed at a pressure of 900 MPa had a Knoop hardness of 90 Kg/mm^2, approximately equal to that of annealed aluminum. In comparison, MoS_2 pellets formed in the same way had a Knoop hardness of 13 Kg/mm^2. With its amorphous

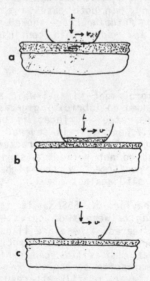

Fig. 12. Idealized models describing sliding of two bodies with an intervening film. (a) Film adheres to both surfaces. Sliding is accomplished by intrafilm flow. (b) Films are adherent to both bodies. Sliding is interfilm. (c) Film does not adhere to bodies. Sliding is at inter-face.

structure there is no mechanism by which $SbSbS_4$ can shear easily other than by viscous flow brought about by thermal softening which was not observed. $SbSbS_4$ when applied as a dry powder to sliding steel contacts formed a thick, tenatious layer that appeared to accommodate sliding by a destructive fracturing-compaction process. Although the steel surfaces were considerably harder than $SbSbS_4$, they sustained wear as a result of a chemical reaction process coupled with the mechanical stresses associated with sliding.

When used as a pellet, $SbSbS_4$ exhibited an even higher coefficient of friction than as a powder. Because of the strong adhesion and the poor flow characteristics of $SbSbS_4$, wear was rapid. Wear may also have been enhanced by the adherent patches of film which could plow the pellet surface. Plowing appeared to be a major cause of wear of the MoS_2 pellet. In that case a thick film of considerable surface roughness was formed on the ring. The behavior of Sb_2S_3 and FeS_2 pellets was similar to that of MoS_2. The fact that the friction coefficient of these materials was higher than MoS_2 can be attributed to the greater strength of the pellets. $Fe_{0.9}S$ differed from the other materials studied in that a thin, uniform film was formed on the ring surface and pellet wear was small. The friction coefficient was in the same range as Sb_2S_3 and FeS_2 but did not exhibit the wide fluctuations. $Fe_{0.9}S$ is apparently a material (at least when in the pellet form) that possess a combination of strength, shear, and adhesion properties that yields smooth sliding without high wear. It is interesting that this form of iron sulfide was identified on specimen surfaces lubricated with grease containing $SbSbS_4$ (5).

CONCLUSIONS

1. When $SbSbS_4$ is used as a dry powder to lubricate steel, friction and wear are high.
2. High friction is associated with the strong adhesion of $SbSbS_4$ to steel, its amorphous structure, and its relatively high shear strength.
3. A chemical reaction between $SbSbS_4$ and steel to form FeS_2 is a contributing factor to the high wear rate.

Acknowledgement: This work was supported by the Office of Naval Research for which the authors are grateful.

References

1. Devine, M.J., Cerini, J.P., Chappell, W.H., and Soulen, J.R., ASLE Transactions 11, 283-289 (1968).
2. Niazy, S.M., Chappell, W.H., and Soulen, J.R., NLGI Spokesman 35, No. 12, 420-426 (1972).
3. King, J.P. and Asmerom, Y., ASLE Transactions 24, 497-504 (1981).
4. Ives, L.K., Peterson, M.B., Harris, J.S., Boyer, P.A., and Ruff, A.W., Investigation of the Lubrication Mechanisms of the Complex Metal Sulfide, SbSbS4, Nat. Bur. Stand. (U.S.) Spec. Publ. 82-2545 (1982).
5. Ives., L.K., Harris, J.S., and Peterson, M.B., Wear of Materials 1983, The American Society of Mechanical Engineers, New York, (1983), pp. 507-513.
6. ASTM Standard D2714, 1981 Annual Book of ASTM Standards, Part 24, American Society for Testing Materials, Philadelphia, PA, (1981).
7. Tanaka, K., Uchiyama, Y., Makagawa, T., and Matsunaga, M., Wear of Materials 1981, The American Society of Mechanical Engineers, New York, (1981), pp. 637-643.
8. Matsunaga, M., Nakagawa, T., and Tennichi, M., ASLE Transactions 26, 64-68 (1983).
9. Bridgman, P.W., Proc. Amer. Acad. Arts and Sci. 71, 387-460 (1936).

APPENDIX

MFPG Publications

Both printed and microfiche copies of the following MFPG publications (1-12) whose catalog numbers start with either "AD" or "COM" may be obtained from the NTIS.

> National Technical Information Service
> 5285 Port Royal Road
> Springfield, Virginia 22151

1. Glossary of Terms . . . . AD 721 354

2. Proceedings of Meeting Nos. 1-9 (set of 5) . . . . AD 721 359

   Meeting Nos. 1-5   Papers and discussion on Failure
                       Analysis and Control

   Meeting No. 6   "Detection, Diagnosis and Prognosis"
                       December 6, 1968

   Meeting No. 7   "Failure Mechanisms as Identified with
                       Helicopter Transmissions"
                       March 27, 1969

   Meeting No. 8   "Critical Failure Problem Areas in the
                       Aircraft Gas Turbine Engine"
                       June 25-26, 1969

   Meeting No. 9   "Potential for Reduction of Mechanical
                       Failure through Design Methodology"
                       November 5-6, 1969

3. Proceedings of Meeting No. 10 . . . . AD 721 912
   "Vibration Analysis Systems"
   January 21-22, 1970

4. Proceedings of Meeting No. 11 . . . . AD 724 475
   "Failure Mechanisms: Fatigue"
   April 7-8, 1970

5. Proceedings of Meeting No. 12 . . . . AD 721 913
   "Identification and Prevention of Mechanical
   Failures in Internal Combustion Engines"
   July 8-9, 1970

6. Proceedings of Meeting No. 13 . . . . AD 724 637
   "Standards as a Design Tool in Surface Specifi-
   cation for Mechanical Components and Structures"
   October 19-20, 1970

7. Proceedings of Meeting No. 14 . . . . AD 721 355
   "Advances in Decision-Making Processes in
   Detection, Diagnosis and Prognosis"
   January 25-26, 1971

8. Proceedings of Meeting No. 15 . . . . AD 725 200
   "Failure Mechanisms: Corrosion"
   April 14-15, 1971

9. Proceedings of Meeting No. 16 . . . . AD 738 855
   "Mechanical Failure Prevention through
   Lubricating Oil Analysis"
   November 2-4, 1971

10. Proceedings of Meeting No. 17 . . . . AD 750 411
    "Effects of Environment upon Mechanical
    Failures, Mechanisms and Detection"
    April 25-27, 1972

11. Proceedings of Meeting No. 18 . . . . AD 772 082
    "Detection, Diagnosis and Prognosis"
    November 8-10, 1972

12. Proceedings of Meeting No. 19 (NBS SP 394) COM-74-50523
    "The Role of Cavitation in Mechanical
    Failures"
    October 31-November 2, 1973

Printed copies of the following MFPG publications may be obtained from the U. S.
Government Printing Office.

        Superintendent of Documents
        U. S. Government Printing Office
        Washington, DC 20402

Microfiche copies of these publications may be obtained from the NTIS.

13. Proceedings of Meeting No. 20 (NBS SP 423) SN003-003-01451-6
    "Mechanical Failure - Definition of
    the Problem"
    May 8-10, 1974

14. Proceedings of Meeting No. 21 (NBS SP 433) SN003-003-01639-0
    "Success by Design: Progress through
    Failure Analysis"
    November 7-8, 1974

15. Proceedings of Meeting No. 22 (NBS SP 436) SN003-003-01556-3
    "Detection, Diagnosis and Prognosis"
    April 23-25, 1975

16. Proceedings of Meeting No. 23 (NBS SP 452) SN003-003-01664-1
    "The Role of Coatings in the Prevention
    of Mechanical Failures"
    October 29-31, 1975

17. Proceedings of Meeting No. 24 (NBS SP 468) SN003-003-01760-4
    "Prevention of Failures in Coal
    Conversion Systems"
    April 21-24, 1976

18. Proceedings of Meeting No. 25 (NBS SP 487) SN003-003-01829-5
    "Engineering Design"
    November 3-5, 1976

19. Proceedings of Meeting No. 26 (NBS SP 494) SN003-003-01844-9
    "Detection, Diagnosis and Prognosis"
    May 17-19, 1977

20. Proceedings of Meeting No. 27 (NBS SP 514) SN003-003-01935-6
    "Product Durability and Life"
    November 1-3, 1977

21. Proceedings of Meeting No. 28 (NBS SP 547) SN003-003-02083-4
    "Detection, Diagnosis and Prognosis"
    November 28-30, 1978

22. Proceedings of Meeting No. 29 (NBS SP 563) SN003-003-02120-2
    "Advanced Composites"
    May 23-25, 1979

23. Proceedings of Meeting No. 30 (NBS SP 584) SN003-003-02272-1
    "Joint Conference on Measurements and
    Standards for Recycled Oil/Systems
    Performance and Durability"
    October 23-26, 1979

24. Proceedings of Meeting No. 31 (NBS SP 621) SN003-003-02428-7
    "Failure Prevention in Ground
    Transportation Systems"
    April 22-24, 1980

25. Proceedings of Meeting No. 32 (NBS SP 622) SN003-003-02361-2
    "Detection, Diagnosis and Prognosis:
    Contribution to the Energy Challenge"
    October 7-9, 1980

26. Proceedings of Meeting No. 33 (NBS SP 640) SN003-003-02425-2
    "Innovation for Maintenance
    Technology Improvements"
    April 21-23, 1981

27. Proceedings of Meeting No. 34 (NBS SP 652) SN003-003-02488-1
    "Damage Prevention in the
    Transportation Environment"
    October 21-23, 1981

Printed copies of the following MFPG publications are available from Cambridge
University Press, 32 East 57th Street, New York, NY 10022.

28. Proceedings of Meeting No. 35
    "Time-Dependent Failure Mechanisms
    and Assessment Methodologies"
    April 20-22, 1982

29. Proceedings of Meeting No. 36
    "Technology Advances in Engineering and
    Their Impact on Detection, Diagnosis
    and Prognosis Methods"
    December 6-10, 1982

226